Forest Quality

Forest Quality
Assessing Forests at a Landscape Scale

*Nigel Dudley, Rodolphe Schlaepfer, William Jackson,
Jean-Paul Jeanrenaud and Sue Stolton*

earthscan
from Routledge

First published by Earthscan in the UK and USA in 2006

For a full list of publications please contact:

Earthscan
2 Park Square, Milton Park, Abingdon, Oxfordshire OX14 4RN
711 Third Avenue, New York, NY 10017

First issued in paperback 2016

Earthscan is an imprint of the Taylor & Francis Group, an informa business

ISBN 13: 978-1-138-97451-7 (pbk)
ISBN 13: 978-1-84407-278-1 (hbk)

Typesetting by JS Typesetting Ltd, Porthcawl, Mid Glamorgan
Cover design by Susanne Harris

A catalogue record for this book is available from the British Library

Library of Congress Cataloging-in-Publication Data has been applied for.

Contents

Part 4 Appendices: Broader Issues and Sources of Information

List of Figures, Tables and Boxes

Figures

Tables

Boxes

Acknowledgements

The forest quality concept has been developed over several years and has gained from the advice and input of many people around the world. The work has been coordinated by the authors of the book, but we wish to stress that it has drawn on the ideas and opinions of numerous people who have shaped and developed the concepts in workshops, field visits, comments on drafts and conversations. So many people have been involved that it is difficult to be certain of listing everyone and we apologize for any inadvertent omissions. The following are to be thanked for their help (although they are not responsible for remaining errors of fact or opinion).

'Wale Adeleke (Nigeria) · Mark Aldrich (WWF International, Switzerland) · Andrew Allo Allo (formerly WWF, Cameroon) · Per Angelstam (Grimso Research Station, Sweden) · Martin Ashby (formerly Common Lands Survey, Wales, UK) · Ndinga Assitou (formerly IUCN, Cameroon) · Joseph Bakakoula (WWF, Cameroon) · Michael Baltzer (WWF, Danube-Carpathian Programme) · Clare Barden (UK) · Garo Batmanian (formerly WWF, Brazil) · Martin Bemelingue (WWF, Cameroon) · Nurit Bensusan (formerly WWF, Brazil) · Wim Bergmans (Netherlands Committee for IUCN, Amsterdam) · Jill Bowling (WWF, UK) · Guido Broekhoven (IUCN, South East Asia) · Dirk Bryant (formerly World Resources Institute, US, now The Nature Conservancy) · Gerardo Budowski (University of Peace San José, Costa Rica) · Rita Bütler (EPFL, Lausanne) · Jane Carter-Lengeler (Inter-co-operation, Switzerland) · Sue Clifford (Common Ground, UK) · Geoff Davidson (Singapore) · Dominick DellaSala (WWF, US) · Mark Dillenbeck (formerly with IUCN) · Mary Edwards (Southampton University, UK) · Chris Elliott (WWF, Switzerland) · J Elmberg (Sweden) · J Everard (UK Forestry Commission) · Malcolm Fergusson (Institute for European Environmental Policy, UK) · Christian Gamborg (Denmark) · Graham Gill (Forestry Commission, UK) · Don Gilmour (Brisbane, Australia) · Christian Glenz (EPFL) · Robin Grove-White (Lancaster University, UK) · Arlin Hackman (WWF, Canada) · Chris Hails (WWF International, Switzerland) · Elie Hakizumwmi (formerly IUCN Cameroon, now WWF) · Pierre Hauselmann (Puly, Switzerland) · Lisa Hooper (Countryside Commission, UK) · Steve Howard (formerly WWF International, UK) · John Innes (University of British Columbia) · Sally Jeanrenaud (IUCN, Switzerland) · Julia Allen Jones (Oregon State University, US) · Scott Jones (Litchfield, UK) · Calestous Juma (Harvard University, US) · Tony Juniper (Friends of the Earth, UK) · Valerie Kapos (UNEP-World Conservation Monitoring Centre, UK) · Lassi Karivalo (Natural Heritage Services Helsinki, Finland) · Harri Karjalainen (WWF, Finland) · Keith Kirby (English Nature, UK) · Alex Korotkov (UNECE, Switzerland) · Henry Lamb (University College of Wales, Aberystwyth, UK) · Heiko Liedeker (Forest Stewardship Council) · Anders Lindhe (WWF) · Stephanie Mansourian (Switzerland) · Martin Mathers (formerly WWF, Scotland) · Alan Mattingley (formerly Ramblers Association, UK) · Tom McShane (Switzerland) · Bihini Won Wa Musiti (IUCN, Cameroon) · François Nectoux (Thames University, UK) · Paulinus Ngeh (Yaoundé, Cameroon) · Nguyen Cu (Viet Nam) · Nguyen Thi Dao (Seychelles) · John Palmer (Tropical Forestry Services Ltd, UK) · Tim Peck (Switzerland) · George Peterken (UK) · Borje Pettersson (Stora-Enso, Sweden) · Duncan Poore (UK) · Kit Prins (UNECE, Switzerland) · Michael Rae (formerly WWF, Australia) · Simon Reitbergen (IUCN, Switzerland) · Per Rosenberg (WWF International) · Ugis Rotbergs (WWF, Latvia) · Sarah Russell (Geneva) · Meri Saarnilahti-Becker (Helsinki, Finland) · Alberto Salas (IUCN, Costa Rica) · Jukka Salo (University of Turku, Finland) · Jeffrey Sayer (WWF International) · Erik Sollander (Jongkoping, Sweden) · Ian Standing (Forest of Dean Information Centre, UK) · Kerstin Stendahl-Rechardt (Finland) · Andy Tickle (CPRE, UK) · Florangel Villegas (IUCN, Costa Rica) · Hardy Vogtmann (BNF, Germany) · Vuong Tien Manh (FPD, Viet Nam) · Lawrence Woodward (Elm Farm Research Centre, UK) · Mark York (Dolgellau, UK) · Olle Zachrison (Umea, Sweden).

The concept was first suggested by Sue Stolton in 1992 and developed by Sue and Nigel Dudley in a project for WWF managed by Chris Elliott. Some of the ideas drew originally on concepts discussed at a colloquium on food quality run jointly by the University of Kassel and Elm Farm Research Centre at Newbury in the UK.

They were developed in more detail at a workshop in the Forest of Dean, arranged by Jean-Paul Jeanrenaud of WWF UK in association with the UK Forestry Commission and with representation from many government and non-governmental organizations (NGOs); the ideas were substantially refined at that time. Further refinement of the concepts took place in a paper by Nigel Dudley and Chris Elliott for an experts' meeting on forest resources assessment, held in Kotka, Finland in 1995. Comparison of indicators of forest quality was initially carried out in a paper prepared by Nigel Dudley for the Convention on Biological Diversity for use at a workshop in Helsinki and was further refined at a follow-up meeting in Wageningen in The Netherlands. Work on biodiversity was carried out for a paper by Nigel Dudley and Jean-Paul Jeanrenaud, presented by Jean-Paul at a conference organized by the European Forests Institute in Monte Verita, Switzerland in 1996. The structure of the introductory paper comes from a lecture given by Nigel Dudley at the science museum in Barcelona at the invitation of Enciclopédia Catalana in 1996 to launch a book in their *Biosfera* series.

Further development took place during 1998. William Jackson suggested the concept of levels of assessment and accompanying diagram at a workshop in Yaoundé, Cameroon in 1998 and Guido Broekhoven and 'Wale Adeleke suggested expanding the indicators of authenticity at the same time. Rodolphe Schlaepfer, Jean-Paul Jeanrenaud and Nigel Dudley had a working trip in the Jura forests of Switzerland in the summer of 1998 to further refine the concepts. Rodolphe Schlaepfer and Jill Bowling contributed substantially to the development of the definition of forest quality, and Rodolphe and John Innes also suggested conflating two criteria on authenticity and forest health at a workshop in April 1998 in Geneva. Jane Carter-Lengeler and Sally Jeanrenaud added several indicators to the social and environmental indicators at that time and Per Angelstam made suggestions on the question of landscape. Sue Stolton and Nigel Dudley initially developed assessment ideas in the Snowdonia National Park, Wales in June 1998. Sue Stolton, Nigel Dudley and Martin Ashby carried out an initial test of rapid forest quality assessment in the Dyfi Valley, mid-Wales in March 1999.

Guidance came during a number of meetings and workshops. These include: the Experts' Meeting on Criteria and Indicators on Sustainable Forest Management, Montréal, October 1993; a joint WCMC/WRI workshop in Cambridge, UK in 1994; a WWF conference in Zvolen, Slovakia in 1995; a presentation in Riga, Latvia in 1994; a European Commission Cost Action meeting in Fontainebleau, France in 1996; a Global Biodiversity Forum meeting in Washington, DC in 1997; the Experts' Meeting on Criteria and Indicators for the CBD in Helsinki, Finland in 1997; and the workshops that have taken place during the project.

The project managers are more than grateful to WWF and latterly to IUCN for both financial and moral support over the years. Substantial development was also possible through the generosity of the late Mr Nicola of Vaud, Switzerland.

Preface

The following manual is a practitioners' guide assessing forest quality at a landscape scale.

The book describes a *framework for forest quality assessment* that can be tailored to individual needs and to a range of outputs. It summarizes work by the World Wide Fund for Nature (WWF), The World Conservation Union (IUCN) and the École Polytechnique Fédérale de Lausannne, in association with the German development organization, Deutsche Gesellschaft für Technische Zusammenarbeit (GTZ), including field-testing in Europe, Central America and the Congo Basin in Africa. The framework aims to provide information for a number of distinct purposes:

- identifying the current and future potential of forested landscapes from environmental and social perspectives;
- distinguishing between different levels of ecological forest quality at a landscape scale to aid in prioritizing conservation interventions;
- planning conservation interventions within priority landscapes identified in ecoregional planning processes or similar;
- providing a basis for negotiation about trade-offs between different forest uses and development of a vision for a forest landscape;
- developing a monitoring and evaluation framework for a variety of conservation actions – protection-management-restoration – within a landscape;
- assessing specific elements of forest quality as part of wider research;
- undertaking long-term monitoring of conditions within a forested landscape.

Application of the framework can vary from being a first, coarse and approximate assessment of conditions to a detailed research programme. It can also be used to provide a single 'snapshot' in time, an indication of trends or long-term monitoring of progress over time. Examples of different uses are included in the book.

Although developed for use in the forest sector, the thinking behind the approach could equally be applied to the assessment of other natural and cultural resources such as marine ecosystems, freshwater and more generally to assessment of landscape or seascape values.

List of Acronyms and Abbreviations

C&I	criteria and indicators
CBD	Convention on Biological Diversity
CEC	Commission for the European Communities
CIFOR	Center for International Forestry Research
DEVP	Dyfi Eco Valley Partnership
FAO	Food and Agriculture Organization
FSC	Forest Stewardship Council
GIS	geographical information systems
GTZ	Deutsche Gesellschaft für Technische Zusammenarbeit (German technical development organization)
HCVF	High Conservation Value Forests
HEP	hydroelectric power
ISO	International Organization for Standardization
ITTO	International Tropical Timber Organisation
IUCN	The World Conservation Union
MAB	Man and the Biosphere
MCPFE	Ministerial Conference for the Protection of Forests in Europe
MINEF	Ministére des forêts et de la faune
NGO	non-governmental organization
NTFP	non-timber forest product
NWGS	non-wood goods and services
PEFC	Programme for the Endorsement of Forest Certification
PES	payment for environmental services
PRA	participatory rural appraisal
P&C	principles and criteria
RAPPAM	Rapid Assessment and Prioritization of Protected Area Management
SWOT	strengths, weaknesses, opportunities and threats
UNECE	United Nations Economic Commission for Europe
UNESCO	United Nations Educational, Scientific and Cultural Organization
WDPA	World Database on Protected Areas
WRI	World Resources Institute
WWF	World Wide Fund for Nature

Part 1
Measuring Forest Quality

1 | What is Forest Quality?

The Tree that moves some to tears of joy is in the Eyes of others only a Green thing that stands in the way

The poet and artist William Blake, circa 1800 in a letter to Reverend John Trusler

In the foothills of the Snowdonia National Park, in Wales, we're looking for an abandoned village; a scatter of houses left behind when a slate quarry closed. It is marked on the map and clearly visible from the road, but is nowadays surrounded by a dense sward of conifers in one of the state-owned forests. We are probably trespassing, forcing our way up a steep slope – often literally on our hands and knees – through dense stands of Sitka spruce. No-one can have been in here for years. The ground is covered with a thick mat of needles, empty of any plants except for the odd place where a windfall has created a little pool of light and life; and here the burst of green forms a sudden contrast with the featureless surroundings. The dense foliage muffles sounds as well, so that we are in virtual silence. It's peaceful, but rather unreal.

The few ruins, when we finally reach them, seem as remote as a village hidden away in a tropical jungle and it is hard to imagine this place as it must have been 50 years ago, stuck out on the edge of a bare hillside, with quarrymen patiently cutting slates for roofing. Most of the men died young, their lungs clogged with decades' worth of thick dust. This operation was obviously abandoned in a hurry. There are still hundreds of slates stacked neatly as if ready for sale, although they are now frost-shattered and covered in a thick growth of lichen, and all the cottages have lost their roofs.

We take a different route out, slithering uncomfortably down a slope where tree branches pull at our faces and hair, but then suddenly burst out into a scrap of remnant oak woodland left around the banks of a stream. The change is immediate, like switching on a light in a darkened room, a burst of new colours and sounds. There is a range of trees: sessile oaks interspersed with birch,

Note: In the Snowdonia National Park, Wales, ancient native woodlands and exotic conifer plantations are both labelled 'forests' but their qualities are very different.

Source: Nigel Dudley

FIGURE 1.1 **Two views of Snowdonia**

hazel and yew. Twisted tree branches are dripping with lichens and mosses, and we have to clamber over fallen logs, while underfoot there is a rich profusion of flowers and ferns. The trees are full of birdsong and overhead the mewing call of a buzzard sounds above the canopy.

The two worlds, pressed up against each other physically, remain in other ways a universe apart. Yet both the conifer plantation and the oak wood are 'forests'. And they both have their uses and their champions. At the heart of this book lies an attempt to understand the differences between the quality of different forests – many far more subtle than the deliberately stark example from Wales – and what 'forest quality' means in practice.

Quantity and quality

Everybody knows that the world is losing forests – images of deforestation fill our magazines and television screens. But it is not just the number of trees that matters; the quality of the forest is also important. Even where the forest area is stable or increasing, there are often rapid changes in its character. Natural forests are being replaced by plantations or by intensively managed forests. Forests around the world are generally becoming younger and less diverse, in both species and structure; this has important impacts for biodiversity and also affects many human values.

A tree plantation is as different from a natural forest as a football pitch is from a wildflower meadow: both may have their place in the forested landscape but it is important that we distinguish between them and understand their different qualities.

Throughout the 1980s and early 1990s, global concerns about forest conservation focused on the rapid rate of deforestation in tropical countries. While this is a real and continuing issue, it is only one half of a more complex problem of global forest management. Growing interest in the status of temperate and boreal forests resulted in recognition of the importance of social or ecological values. Forest *quality* was recognized to be as important an issue as the *quantity* of forest remaining (Dudley, 1992).

There is a growing perception that global forest quality is declining as a result of human activities. Ecologists have become concerned about the replacement of biologically rich old-growth forest with species-poor young forests, intensively managed forests or plantations and the decline in the health of trees and other forest species as a result of anthropogenic changes, especially air pollution and climate change, but also as a result of introduced pests and diseases and invasive species. This in turn has led to a breakdown in the ecological support systems associated with forests including hydrological systems, soil structure and fire ecology. People interested in social welfare and development complain about threats to social rights in forests including issues related to tenure, access and changes of management that have resulted in a decline in non-wood goods and services (NWGS). Lastly and more generally, the changes are resulting in more intangible and hard-to-measure losses to the aesthetic, cultural and spiritual values that many people demand from forests.

Each of these aspects of 'forest quality' has its own champions and detractors. The public debate about the role of both plantations (Carrere and Lohmann, 1996) and air pollution (Dudley et al, 1985), for instance, has frequently been bitter. In those countries where forest cover has stabilized – particularly in the richer temperate and boreal countries of Europe, the Commonwealth of Independent States, the US, Japan, Australia and New Zealand – the debate about forests has shifted from how *much* forest we need to what *kind* of forests remain or could be recreated.

Although the discussion about forest quality initially centred on temperate and boreal forests, as it gained attention, concern about quality has spread to tropical areas as well. The focus of conservationists working in the tropics has been on conserving remaining areas of primary forest. A sharp distinction has been made – at least in theory – between 'natural forest' and 'disturbed forest', although these categories are often poorly defined. Forest that has been disturbed or selectively logged is frequently relegated to a low status in terms of its conservation value. Indeed, it is sometimes not referred to as 'forest' at all; it is for example not uncommon to hear conservationists say that a country like Cameroon has 'barely any forest left at all', even though around 20 million hectares of the country is covered by predominantly natural forest vegetation (Global Forest Watch, 2000). However, this distinction is becoming increasingly hard to maintain as more and more areas of apparently remote tropical forests are also disturbed – sometimes dramatically. A research study published by the German technical development organization Deutsche Gesellschaft für Technische Zusammenarbeit (GTZ) estimated that 32 per cent of forests in the tropics are 'secondary' even using a fairly coarse definition of secondary as *open forest, long fallow and fragmented forest* (Emrich et al, 2000). A more precise definition, including all forests where disturbance has taken place in the recent past, would include a much larger proportion of the total.

But what exactly is forest quality?

Quality means different things to different people. Commercial timber producers will probably not look at a forest in the same way as local villagers, holidaymakers or indigenous people. Yet their views are all valid. The needs of wild plants and animals may not always be the same as our own. Forests give us an astonishing range of goods and services, and reconciling these within a policy of sustainable forest management presents a major challenge to planners and managers. Some countries, such as Germany, have attempted to achieve this by managing forests so that each particular forest stand supplies a wide range of economic, social and environmental benefits, while countries like New Zealand have made a sharp distinction between commercial timber and fibre plantations and 'natural' forests managed for biodiversity and social values.

In practice, some qualities are hard to reconcile: for example timber production and wilderness values. Many forests that are supposedly managed for multiple purposes ('multipurpose forests') have tended to exclude or underplay certain values. However, although a single forest stand cannot easily supply all the potential forest goods and services, this should be possible in a well-designed and managed forest *landscape*, containing a mosaic of different land uses. For example, some forests might be set aside particularly for specialized needs like biodiversity conservation, watershed protection or wood production, while others will serve a range of different functions. We are interested here principally in forest quality on a landscape level; that is, in the overall values of many different areas of forest within one landscape mosaic.

To create forest landscapes that serve many requirements, we need to understand what makes up forest quality, both for wildlife and for people: to understand that, we need first to understand and to cater for different perceptions of forest quality. This is at the core of the forest quality project and the framework for assessment described in this book.

A brief overview of changes in global forest quality

This book is principally about assessment, but a brief discussion of how forests changed during the last century might help set the scene for what follows.

Generally, forests have declined in naturalness over the last 100 years. In some areas, such as western Europe, Japan and much of eastern North America, natural forests were largely cleared hundreds or thousands of years ago, and here the change was more in an increasing 'standardization' of secondary forests. Research undertaken by the United Nations Economic Commission for Europe (UNECE) found that most European countries have less than 1 per cent of their forests in anything approaching a natural state (FAO and UNECE, 2000). National correspondents were asked to estimate the area of 'forest undisturbed by man' as an approximation of 'naturalness', which was defined as forests that had no human disturbance or had been disturbed so long ago that natural processes were completely re-established. According to replies received, 55 per cent of forest studied by the

Temperate and Boreal Forest Resource Assessment can be classified as 'natural'. However, this global figure is distorted by the forest rich areas of Canada and the Russian Federation, and outside these countries the figure for forest drops to just 7 per cent of the total, with most of this in the US and Australia. Sweden records 16 per cent of its forest as natural, Finland 5 per cent and Norway 2 per cent. In the rest of Europe the proportion is usually from zero to less than 1 per cent (Dudley and Stolton, 2004).

Similar changes are now taking place in tropical forests. Although most tropical forests have also long been affected by human activity (Posey and Balee, 1989), until recently this has often been relatively subtle and tropical forests have in general retained a far more natural ecology and structure. This is now changing. Forest degradation affects many of the tropical wet and dry forests that remain, most commonly through logging out the largest individuals or changing forests as a result of overgrazing, unsustainable harvest of non-timber forest products (NTFP), changes to fire regimes and fuelwood collection.

These changes have had a marked impact on biodiversity. Consistent analyses over the last 20 years have found the highest levels of threat to terrestrial species being amongst those found in forests: this is true both for species in developing countries and in highly developed countries. Analysis of the 2000 IUCN Red List of Threatened Species, found that 74 per cent of threatened bird species are almost entirely confined to a single habitat and of these, 75 per cent are dependent on forests (though in each case figures refer to that proportion of threatened species where analysis was possible). Tropical forests contain a high proportion of the threatened species, including 900 bird species. In addition 33 per cent of threatened mammals use lowland tropical rainforest and 22 per cent use montane tropical rainforest. Habitat loss is the over-riding threat to wildlife including for example 89 per cent of threatened birds, 83 per cent of threatened mammals and 91 per cent of threatened plants (analysis focused mainly on trees), and selective logging alone threatens 31 per cent of threatened bird species (Dudley and Mansourian, 2003 drawing on the IUCN Red Data List). In Finland, one of the countries with the highest proportion of forest cover in the world, 44 per cent of the almost 1700 species listed in the Finnish Red Data Book are associated with forests.

At the same time, there has been increased recognition of the value of forests in terms of their environmental benefits, principally through their value in protecting watersheds to supply high quality drinking water, their role in soil control and prevention of avalanches and their potential to sequester carbon. For instance, roughly a third of the world's 100 largest cities draw a significant proportion of their drinking water from forests within protected areas, and protection has often been spurred by recognition of their value in maintaining high quality water (Dudley and Stolton, 2003a). Forests are also proven barriers to erosion. Many of the earliest successful attempts at reforestation, in Austria, Japan and Switzerland, were spurred by concern about rapid soil erosion and catastrophic avalanche damage (Küchli et al, 1998). A number of countries have identified various types of 'conserved forests' to classify these areas, and the concept that states set aside areas of land specifically for their environmental services is now widely accepted. International initiatives, such as the Convention to Combat Desertification and the Convention on Biological Diversity, explicitly recognize the importance of forests from the perspective of environmental management.

Forest quality has also changed from social and economic perspectives. Most commercial attempts to manage forests have focused primarily on timber and fibre, and indeed the increased efficiency of forests as producers of valuable raw materials has been a major driver behind the changes in the quality of the forests that remain. Increasing use of monocultures, including of genetically similar stock, and of intensive management within secondary forests has dramatic impacts on the structure and the ecosystem functioning of forests and also changes their appearance. Fears that the world would run out of timber have proved premature and most recent analyses conclude that supply is likely to meet or exceed demand (Nilsson, 1996; Solberg et al, 1996; Sedjo, 1999; Victor and Ausubel, 2000) although the impacts of climate change are unpredictable.

Other more minor economic and social uses of forests have suffered declines, as timber production has reduced the availability of some NTFPs and as local and indigenous peoples have lost their traditional forest homes to agricultural and timber plantations, cattle ranching and protected areas. In particular, forests have been expropriated from indigenous people, who have often had little chance to exercise any land rights. A report from the United Nations Research Institute for Social Development concludes that: 'Land tenure problems are often root causes – or play an important mediating role – in deforestation...' (Dorner and Thiesenhusen, 1992).

In general then, forests in many parts of the world have become less natural and more narrowly focused on the production of timber and fibre (Matthews et al, 2000). Over the past two decades there has been something of a backlash against such changes and an increasing effort to integrate forest management with other forest values, including an emphasis on forests' role in recreation, cultural survival and spiritual values. The introduction of sustainable management initiatives and the certification of forest management under the auspices of organizations such as the Forest Stewardship Council and the Programme for the Endorsement of Forest Certification schemes (PEFC) have created an atmosphere in which forest managers see their role increasingly as land stewards rather than simply timber producers (Garforth and Dudley, 2003).

Addressing wider needs from forests implies a better understanding of exactly what it is that forests provide and the starting point for the work on assessment of forest quality.

Assessing forest quality

The concept of forest quality was first developed as a lobbying and communication tool for WWF International (formerly known as the World Wide Fund for Nature) and The World Conservation Union (IUCN), to outline simply and clearly the range of values that the organizations believed to be important in forests. Because it dealt with *all* forest values, it also helped the sponsoring conservation organizations to re-examine their own priorities, particularly with respect to the links between human societies and forests.

The word *quality* was chosen because it is generally seen as a positive and powerful term, focusing on benefits and opportunities. It can include a range of ecological, social and economic values and translates easily into a range of other languages. It is also, by its definition, a word that requires explanation and interpretation and below we give our own definition of 'forest quality' underpinning the assessment method that forms the core of this book.

Looking at forest quality has helped us look again at our own values and assumptions: sometimes it has meant revising our own opinions about certain forms of forest management or approaches to forests. It is hoped that this book will help other people do the same.

A definition of forest quality

'Quality' is a value-laden word – things can have both positive and negative quality, although used alone the word often implies a positive value. More importantly, perceptions of quality and the values assigned to a particular quality differ from one person to another. Perceptions are also changed by the range and depth of the observer's knowledge; casual onlookers may well miss subtle quality values such as microhabitats for rare species, particular timber values or the spiritual importance that people assign to individual tree groves or landscape features within the forest. Other values, including biodiversity richness, are often poorly understood in many of the less well-studied forests, making assessment difficult. The tension between these different quality values lies at the heart of efforts to assess forest quality and to make good use of the results as a tool for improving management. More generally, it reflects the difficulty in reaching consensus about the ways in which forests should be managed – where management can range from a decision to leave an area alone to replacing it by agricultural crops or exotic monoculture tree plantations.

For our purposes here, forest quality is defined as: *The significance and value of all ecological, social and economic components of the forest landscape.*

Choosing and categorizing indicators of forest quality

It is impossible to measure the totality of values for any ecosystem, so instead we use a number of *indicators* to give a flavour of the overall quality. For instance, biological surveys cannot practically measure all the species in an area, so they generally select some 'indicator species' that give a picture of the health of plants and animals. Clearly, choosing 'good' indicators is essential.

The indicators used in a forest quality assessment in effect put flesh on the overall definition. The list of indicators used needs to be broad enough to cover as full a range of values as possible, and be relatively easy to collect and clear in the information that they provide. Indicators are discussed in greater detail in Stage 3 of the assessment process presented in Chapter 4.

Most assessment systems, or monitoring systems, categorize or divide indicators into broad groups of *criteria* for ease of use and application, and we do the same here. The ways in which indicators are divided is important in that they help to clarify the way in which they are used and also 'weight' the assessment (for example if an assessment system has several groups of indicators referring to economic issues and only one referring to social issues then it is likely to be biased in favour of economics). The choice of indicators also provides an overall 'philosophy' to the assessment by saying something about the importance of different issues.

In many assessment systems, environment has been relegated to a relatively unimportant element compared with other issues such as economic importance, although there are now also some specialized indicator sets relating to the environment, such as WWF's *Living Planet Index* (Loh, 2003).

Other examples that relate to forests and forest management include:

- the IUCN well-being index that divides indicators into two, relating to human well-being (socio-economic) and environmental (ecological, environmental services etc.) (Prescott-Allen, 2001);
- the Montreal Process criteria and indicators for temperate and boreal countries outside Europe, which uses seven criteria (and 67 indicators) including:
  - conservation of biological diversity;
  - maintenance of productive capacity of forest ecosystems;
  - maintenance of forest ecosystem health and vitality; conservation and maintenance of soil and water resources;
  - maintenance of forest contribution to global carbon cycles;
  - maintenance and enhancement of long-term multiple socio-economic benefits to meet the needs of society;
  - legal, institutional and economic framework for forest conservation and sustainable management.
- The African Timber Organisation that identified 5 principles, 20 criteria and 60 indicators for sustainable forest management.

Many other criteria and indicator processes (see Table 1.1) identify around seven criteria. As these processes are all measuring the same basic values, it is clear that the divisions are fairly arbitrary. At the same time, they do help to define the overall feel and aims of the resulting assessment.

An obvious choice for the forest quality assessment would have been to divide indicators between ecological, social and economic values in line with the definition, and indeed this is an option if users prefer. However, trying to divide indicators along those lines caused us some problems: distinctions were very unclear (many social and ecological values also have clear economic implications) and the divisions created exactly the kind of artificial barriers between forest quality values that we were trying to avoid. Instead, we chose three divisions that we felt made a better attempt at capturing the complexity of forest values:

1 *Authenticity*: captures issues relating principally to ecological values.
2 *Social and economic benefits*: relates to issues primarily affecting human society.
3 *Environmental benefits*: straddles the two, by describing ecological values that also have very direct social and economic implications.

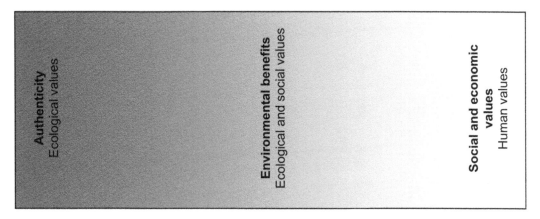

FIGURE 1.2 **Subdividing indicators of forest condition**

The three indicator groups are discussed in greater detail in Table 1.1 below.

TABLE 1.1 **Main groups of indicators of forest quality**

Criteria	Brief description
Authenticity	Authenticity is a measure of ecosystem integrity and health in the broadest sense. It concentrates on current ecosystem functioning, regardless of the forest's history, and thus also has relevance to disturbed forests. A definition of an authentic forest might be a forest in which: *all the expected ecosystem functions can continue to operate indefinitely*
Environmental benefits	Environmental benefits *encompass a range of issues that have direct relevance to both ecosystem health and to the health of human societies.* Important elements include the extent to which forests interact with soil and water systems, the impacts on climate and forests' role in harbouring biodiversity
Social and economic benefits	The last criterion is exclusively focused *on interactions between forests and human societies.* Benefits range from products, such as wood and game, to the use of forests for living or recreation through to hard-to-measure values such as the cultural, aesthetic and spiritual values of particular forest types or locations

In practice, the divisions only become significant if some weighting or scoring is included within the methodology, and can be changed around as conditions or personal preferences dictate

Indeed, each forest landscape is unique and will have to be treated accordingly. In this book we provide a *framework* for how a group or individual might go about assessing forest quality, to be adapted according to circumstances. On the following pages we suggest some useful indicators of forest quality, although these will have to be modified to some extent on a case-by-case basis. The ways in which they are finally divided up will depend to a large extent on what the assessment is aiming to discover.

2 Why Assess Forest Quality at a Landscape Scale?

In a small side room at the United Nations in Geneva, representatives from a group of conservation organizations are hunched around a table in the small hours of the morning, poring over draft text of a resolution on forest management. Over breakfast in a small café in Portland, Oregon, we hear how activists are mounting a campaign to save a particular tract of forest from logging and a few hours later an industry representative complains bitterly about how ill-informed conservationists are destroying working communities throughout the Pacific Northwest. And drifting along a small river through a forest reserve in Sabah, Borneo, we surprise a local man stealthily loading a stolen log onto his dugout canoe.

Decisions about forest management are made at almost every possible scale. Anyone who has been involved in the often exhausting attempts to influence national or international policy does so because they think that decisions made at this level will trickle down all the way to the forest floor and make real differences to the environment and to peoples' lives. Discussions in the boardrooms of major companies can have effects that ripple out over millions of hectares of forest. But in many countries forests are still sculpted much more anarchically, through countless individual actions, often made through necessity or short-term expediency by poor people who have few other options open to them.

In practice then, decisions at all scales are important. This book focuses on a landscape scale and at how we can collect and analyse information about different aspects of forests at this scale. Before jumping into the details of assessment methodologies, this chapter attempts to answer two questions by way of context:

- Why assess?
- Why a landscape scale?

Why assess?

The types of assessments described in this book are undertaken for practical reasons – the results will not just sit gathering dust in an academic journal but will be used to make practical decisions about management. Indeed, carrying out assessments at the scale and about the range of issues required to build up a picture of landscape-level forest quality necessitates cutting corners and making assumptions in ways that professional academics often feel uneasy about. What we are trying to do here is to draw as good an overview as possible, and where necessary be clear about limitations of information. The resulting information can be used for anything from planning projects to monitoring their impacts. The forest quality methodology has been used, for instance, to:

- find out what different stakeholders consider most important in a forest landscape;
- provide information for planning large-scale conservation or development projects;
- assemble information and opinions to help in the negotiation of the trade-offs inevitable in balancing conservation and development priorities on a landscape scale;
- develop monitoring and evaluation systems for measuring success or failure in large-scale projects;
- report on progress in conservation;

In addition, despite the statement that started this section, there is no reason why assessments of forest quality, or particular aspects of quality, should not be used in less applied research, as a way to assemble information quickly for broader-scale ecological or social studies. All the potential uses have in common the need for information that covers both a broad geographical scale and a broad assemblage of information.

Why a landscape scale?

Forest management is attempted at every scale, from global agreements to decisions about whether or not to chop down an individual tree. All scales have their uses. Here we focus on one scale, but before explaining why the landscape is important, we summarize what we already know about working at other scales along with details of how they may be assessed.

FOREST POLICY AT MULTIPLE SCALES

There are no such things as global laws, but there are global treaties and agreements, which in practice carry much the same weight as laws. Countries that regularly break global treaties tend not to have very effective domestic laws either. There is no global forest treaty or convention (and indeed non-governmental organizations (NGOs) have long opposed such a treaty on the grounds that it is likely to weaken existing international agreements) but there are the Forest Principles agreed at the Earth Conservation Summit. More detailed issues relate to decisions within the Convention on Biological Diversity. Other global scale issues relate to principles for when the management of forests to increase carbon sequestration to combat global climate change is eligible for international funding or broad targets for restoration.

At a continental or regional scale, decisions are made through agreements between countries with respect to general principles relating to the amount of protection, levels of productivity and the composition and form of forests, such as the various agreements and criteria and indicators agreed by the Ministerial Conference on the Protection of Forests in Europe and similar regional agreements. Narrowing the scale still further, a number of large NGOs have been promoting the principle of planning at an ecoregional scale, where an ecoregion is defined as 'a large area of land or water that contains a geographically distinct assemblage of natural communities that share a large majority of their species and ecological dynamics, share similar environmental conditions and interact ecologically in ways that are critical to their long term persistence' (Olsen et al, 2001). In total 867 terrestrial ecoregions have been defined around the world.

Many of the day-to-day decisions about approaches to forest management, including identification of forests falling into broad management types such as protected areas, forest reserves or production forests, are taken at national scale. Some countries already have detailed forest inventory systems, some of which go way beyond a simple measure of area of productivity.

ATTEMPTS TO DEFINE FOREST QUALITY AT MULTIPLE SCALES

The 1992 Earth Summit in Rio de Janeiro first set international targets for forest conservation through its *Forest Principles* (UN, 1993). Much derided at the time for being too general and too weak, the Principles did play an important role in setting a precedent for international targets for forest management and these were echoed in time by other, gradually more quantifiable, targets from institutions such as the International Tropical Timber Organisation (ITTO), the Convention on Biological Diversity (CBD) and the United Nations Forum on Forests, as well as a handful of influential targets from NGOs. One implication of the fact that governments had agreed to measurable targets is that they had to make some effort to measure these, and suddenly criteria and indicators of good forest management assumed much higher political importance than before. An early version of the indicators of forest quality described in this manual was published in 1992 (Dudley, 1992) and in more complete form in 1993 (Dudley et al, 1993). During the last decade there have been many attempts to define criteria and indicators (C&I) for sustainable forest management on global, regional, national and forest management unit levels.

Many stemmed from discussions at the UN Conference on Environment and Development (the 'Earth Summit') in 1992 and from the Forest Principles. Most have extended well beyond traditional concerns about growing stock and tree health and include a range of issues relating to environment and human society. Many have also influenced, and have in turn sometimes been influenced by, the C&I of forest quality described here.

At a global level, C&I have been developed for tropical forests by the ITTO and for forest biodiversity by the CBD. Issues of quality were addressed by the temperate and boreal forest component of the Forest Resource Assessment 2000, organized jointly by the Food and Agriculture Organization (FAO) and the United Nations Economic Commission for Europe (UNECE) (FAO and UNECE, 2000). In a parallel process, six other regional criteria and indicator processes have been developed, focusing in greater detail on particular issues relevant to different parts of the world. The first two, the Ministerial Conference on the Protection of Forests in Europe (MCPFE) (formerly known as the Helsinki or Pan European Process) and the Montreal Process, were initiatives of groups of governments; later developments were coordinated by FAO and in one case by the African Timber Organisation. Some of these continue to be influential, while others have since lost funding and appear to have been abandoned. The work at international and regional levels has been complemented by the development of national systems, in part to implement regional C&I processes. There are marked similarities between the regional systems, to the extent that there was an unsuccessful attempt to combine them into one system, associated with an intergovernmental meeting in Helsinki in 1996 (Anttila, 1996). Lastly, there has been a range of non-governmental initiatives. Many are based around stand-level assessments, for instance through forest management certification under the auspices of the Forest Stewardship Council (FSC) or the International Organization for Standardization (ISO). There are also some NGO attempts to reflect forest quality criteria at a regional level, such as the World Wide Fund for Nature (WWF) European forest scorecards (Sollander, 2000). Some initiatives are summarized briefly in Table 2.1.

There have also been a number of attempts to evaluate and compare the various schemes, most notably by the Center for International Forestry Research (CIFOR) (Prabhu et al, 1996) and on a more theoretical basis by the Dutch-based Tropenbos Foundation (van Buren et al, 1997). CIFOR has also developed a methodology for users to select a portfolio of indicators suitable to their particular situation and has tested this in many parts of the world (Prabhu et al, 1999).

The need for a landscape approach to forest quality

'Landscape' is itself a rather nebulous term and open to different interpretations. In this context, we suggest the following definition: 'a contiguous area, intermediate in size between an "ecoregion" and a "site", with a specific set of ecological, cultural and/or socio-economic characteristics distinct from its neighbours'.

A landscape is also a scale at which we have a chance of balancing all aspects of forest quality, which is the reason why it is of interest here. We have already said that a single area of forest cannot supply all the possible goods and services to an optimal degree. Wildlife conservation or recreation activities often do not sit easily with intensive forest production for instance, and certainly not with hunting areas. Faith groups often want their sacred groves to be set apart from other activities. Forests that are valuable for preventing avalanches should not be clear-felled, and so on. Landscapes are valued differently by different stakeholders (Carlson, 1990).

TABLE 2.1 **Examples of previous attempts to define forest quality**

Criteria and indicators (C&I)	Details
Global level processes measuring forest quality on a country scale	
Convention on Biological Diversity (CBD)	The CBD drew up draft C&I for forest biodiversity at an experts' meeting in Helsinki in 1996
International Tropical Timber Organisation (ITTO)	ITTO has drawn up a variety of C&I for tropical forests, including for the conservation of biodiversity (ITTO, undated), natural forest management (ITTO, 1992; ITTC, 1997), plantations (ITTO, 1993), and for forest restoration (ITTO, 2002)
United Nations Forest Resources Assessment	The Forest Resource Assessment 2000, by FAO and the UNECE, included aspects of biodiversity, naturalness and non-timber forest products (Nyyssönen and Ahti, 1996)
Regional level criteria and indicator processes measuring forest quality on a country scale	
Ministerial Conference for the Protection of Forests in Europe (MCPFE)	An initiative launched in 1993, which agreed on a General Declaration and four Forestry Resolutions. The MCPFE has drawn up indicators of good forest management at a national level (MCPFE, 1995; MCPFE, 2002), and used them to report on European forest status (MCPFE Liaison Unit and FAO, 2003)
Montreal Process	Launched in October 1993. It has drawn up criteria and indicators of sustainable forest management with ten non-European temperate and boreal countries and produces regular reports (Canadian Forest Service, undated) including a definition of sustainable forest management (Canadian Council of Foreign Ministers, 1995)
Tarapoto Process	Launched by the Amazon Cooperation Treaty at a Regional Workshop to Define Criteria and Indicators of Sustainability of the Amazon Forest in Peru in February 1995 (Ministry of Foreign Affaris of Peru, 1995)
Dry-Zone Africa Process	Launched at Nairobi in November 1995 (FAO, 1996)
Central American Process	Draft criteria and indicators were developed at Tegucigalpa, Honduras in 1997. C&I are set at regional and national level (CCAD, FAO and CCAB-AP, 1997)
North Africa and the Middle East	FAO process – draft C&I were produced in 1997
African Timber Organisation	Principles and criteria (P&C) for sustainable management of African tropical forests have been developed in cooperation with the ITTO (ITTO, 2003)

TABLE 2.1 **Examples of previous attempts to define forest quality** *(continued)*

Criteria and indicators (C&I)	Details
National level criteria and indicator processes measuring forest quality on a country scale *Many countries now have some systems, developed or under developed, in response to the regional processes outlined above – some examples are given below*	
France	Detailed indicators for French forests (Ministère de l'agriculture et de la péche, 1994)
Finland	C&I for the Pan European Process (Eeronheimo et al, 1997)
Stand-level attempts to set criteria of forest quality *A number of stand-level forest certification schemes overlap with the forest quality criteria. A selection is given below: there are now several competing forest certification systems and individual standards within systems such as the Forest Stewardship Council*	
Forest Stewardship Council (FSC)	An accreditation body for independent, stand-level assessment of sustainable forest management. The FSC Principles and Criteria guide certification bodies, which then draw up their own standards
Programme for the Endorsement of Forest Certification (PEFC)	An alternative certification system, developed by forest owners in Europe
ISO 14000	The International Organization of Standardization (ISO) has developed a certification scheme for timber, although this is not used as an independent assessment at stand level
Soil Association	The organization launched a Responsible Forestry Programme in 1994, associated with the Woodmark label and accredited to the FSC
Center for International Forestry Research (CIFOR)	A series of toolkits developed for choosing and testing criteria and indicators for stand-level forest management, along with specific national criteria and indicators for plantations (Poulsen et al, 2001)
World Resources Institute (WRI)	WRI has carried out a great deal of work on environmental indicators, including indicators of quality (Hammond et al, 1995), sometimes with IUCN (Reid et al, 1993)
The World Conservation Union (IUCN)	IUCN developed a computer software approach to measuring forest well-being using variable indicators (Moiseev et al, 2002)
ProForest	Indicators of High Conservation Value Forest have been developed at stand (Jennings et al, 2003) and landscape level (Jennings and Jarvie, 2003)

We recognize this through establishing a mosaic of management regimes: production forests, plantations, nature reserves and community land for example. From a management perspective, attempts to represent all quality values within a single stand can result in nothing being ideally represented. However, it is possible, with good planning and management, to represent all these values within a forest landscape: for example through a mosaic of natural forests and forests managed for a range of different purposes. In other words, the mosaic should in theory add up to a harmonious whole. The extent to which such a mosaic is made up of areas with multiple functions, or specialized functions, will change depending on individual conditions.

Such insights are not new of course and are reflected in landscape planning exercises in many parts of the world. Very large-scale examples exist, such as the Tennessee Valley River Authority in the US. Many have in the past been quite top-down exercises, although more participatory approaches are starting to emerge. But we still know comparatively little about identifying the right mixture of uses, and still less about how these might be attained, particularly in places where multiple landowners all have their own views about the way to manage their forests. In the current context, increasing recognition of the wide range of forest values also means that these challenges are increasing.

Effective forest planning at a landscape scale, which often also includes a great deal of negotiation if there is more than one landowner, depends on a thorough understanding of what the forest provides. As outlined in Table 2.1 above, almost all existing C&I processes measure forest condition at either a *national level* or at the level of an individual stand or *forest management unit level*. National level C&I are useful for comparing countries and for measuring progress towards sustainability but are too general to show regional variations or to act as a planning tool. Forest management unit level C&I are useful for planning management of an individual forest but say little about how a region is performing, or about whether a stand is being managed appropriately within the wider landscape. For example the existence of FSC certification gives little information about regional protected area networks or about how individual forests contribute to the landscape. The method of forest quality assessment falls between these two extremes, as shown in Figure 2.1.

ECOREGIONAL OR NATIONAL LEVEL

Criteria and indicators (Montreal Process, Tarapoto Process etc.),

Convention on Biological Diversity forest biodiversity indicators, UN Forest Resource Assessment

LANDSCAPE LEVEL
Forest Quality Criteria

FOREST MANAGEMENT UNIT LEVEL

FSC *Principles and Criteria*, ISO-14000, codes of practice

FIGURE 2.1 **Different levels of assessment**

So to put the two questions at the start of this section together: we need assessment systems to help plan and implement good forest management and we need these at a landscape level because this is the scale at which different priorities need to be balanced. A forest quality assessment might be summed up as an attempt to provide meaningful information to make decisions about forest management at a scale that will be understandable to most people. Much of the remaining text describes the approach to assessment, followed by detailed discussion of different indicators and some real-life examples, but before this we look specifically at who might be involved in such assessment systems.

Note: By assessing at the level of the landscape, site characteristics can be put into a wider context, making it easier to judge whether a particular intervention will be positive or negative from social and ecological perspectives.

Source: Nigel Dudley

FIGURE 2.2 **Measuring forest condition in a strict nature reserve in Lithuania**

Who Should Assess Forest Quality?

We are halfway through a public meeting to identify what our research study should focus on. So far a few voices have been dominating the conversation but gradually more people are starting to speak. We already have a huge list of social topics – everyone's pet subject and far more than we can tackle. But there has been a spell of sunny weather and all the farmers are working late on the harvest, so we're missing a whole constituency of interest. Some people seem to have ideas but are too shy to speak. Others veer off at a tangent, but the chair is quite firm and we are gradually building up a picture of local concerns and priorities. We'll have a result that we can work with by the end of the evening but it's still too early to tell just how representative it will be.

We started considering how to assess forest quality from the presumption that assessments would only be valid if they were participatory – that is, if they gave a voice to the full range of stakeholders that lived in or used the forest landscape being studied. But conversely, we also found some good reasons not to follow a participatory approach on occasion and came to the conclusion that the degree and approach to participation needs to be determined virtually on a case by case basis. So in this chapter, we look at arguments for and against using participatory approaches, some of the tools available if you do decide to go down a participatory route and how to make a decision.

Participatory and non-participatory approaches

Developing and testing participatory methods has become a mini-industry amongst development organizations and academics over the last few years. They stemmed in part from recognition that many well-meaning development and conservation projects were missing the very people they were supposed to be helping because those involved had not found out enough about the real needs and desires of local communities. At a more general level, participatory approaches also reflect a more general move towards greater local-level democracy and away from decisions about natural resources being taken by bureaucrats sitting in offices far away. Techniques have evolved particularly to find out the needs and desires of local communities, including simple ways of developing and comparing different options; experience has also developed in local-level negotiation and conflict resolution. Different levels of participation are shown in Figure 3.1.

There are some good reasons for involving people in assessments. If an assessment is being used to make decisions about forest management that will impact on particular groups and communities, then they should generally have the right to input their ideas, knowledge and opinions. This is justifiable from an ethical standpoint but has some practical aspects as well: local communities can often supply information that would be unavailable from any other source, and management changes initiated without taking other perspectives into account often fail.

But there are also some reasonable arguments for caution as well. Assessments are often political and it would be naive to pretend otherwise; immediately involving all stakeholders in a discussion can give powerful vested interests an opportunity to apply pressure for their own ends. For instance an indigenous community wanting to get official protection for a sacred site in a forest might want to gather together some facts before opening up the debate to well-connected timber companies who might wish to cut the whole area down to create a pulp plantation. While transparency is always an ideal, the principle runs into ethically tricky areas if full participation undermines the most vulnerable members of society.

Full control by the agency in charge	Shared control by the agency in charge and other stakeholders				Full control by other stakeholders
	Actively consulting	Seeking consensus	Negotiating (involved in decision-making) and developing specific agreements	Sharing authority and responsibility in a formal way (for example via seats in a management body)	Transferring authority and responsibility
No interference or contribution from other stakeholders					No interference or contribution from the agency in charge

Increasing expectations of stakeholders   ⟶

Increasing contributions, commitment and 'accountability' of stakeholders   ⟶

Source: Adapted from Borrini-Feyerabend, 1996

FIGURE 3.1 **Degrees of participation**

Participation also takes a lot of time. Asking people to give up their time for your research may be justifiable if your research directly affects their livelihoods, but even so it means stakeholders sacrificing time. More academic studies are even more of an imposition. Although there will be occasions when people desperately want to take part in discussions, for example when decisions about management will have important impacts on their livelihoods, on other occasions taking the time to comment on particular issues will be a burden. Generally speaking the less that a community relies directly on natural resources the harder it becomes to encourage people to take part in discussions about their management. In these circumstances interested minorities who have the time to take part can have a disproportionate impact on decisions.

Participatory approaches also have to be presented with care, so as not to raise false expectations. Asking people to spend time giving their views is of little purpose if key decisions have already been taken; it is important to be very clear about exactly what involvement in an assessment process will or will not provide.

Giving the opportunity for participation is often seen as a *privilege* or a *right* 'given' to stakeholders by whoever is initiating an assessment or project. In reality it might better be regarded as a *service* that stakeholders provide. Good assessments need good input and it is up to the assessor to make it as easy as possible for stakeholders to provide this. Selecting a representative group of stakeholders (possibly at random) and paying them for the time needed for them to help an assessment by giving their opinions may be one practical way of addressing these problems as they arise.

Tools for participation

Numerous tools exist to assist participatory processes; given the strong emphasis that has developed on community forest management over the last few years, many of these are aimed directly at forests. Unfortunately from our perspective here, the majority are aimed at individual communities or groups; one of the challenges of landscape-scale forest quality analysis is in scaling up these approaches to use over a wider area and with more people. Some key tools are outlined in Box 3.1, but these are only a fairly small selection of what is available.

Box 3.1 **Tools for participatory approaches**

Participatory Techniques in Community Forestry: A Field Manual – detailed manual outlining and distinguishing between a wide variety of participatory techniques and giving advice about choosing the best approaches for particular situations and how participatory techniques relate to forest management (Jackson and Ingles, 1998).

Where the Power Lies: Multiple Stakeholder Politics over Natural Resources, a Participatory Methods Guide – a four stage process in analyzing, understanding and working with stakeholder groups (Sithole, 2002).

Who Counts Most? Assessing Human Well-Being in Sustainable Forest Management – methodology for determining the most important stakeholders, using determinants such as proximity to forest, pre-existing rights, dependency, poverty and local knowledge (Colfer et al, 1999).

Anticipating Change: Scenarios as a Tool for Adaptive Forest Management – how people can use future scenarios to plan creatively, describing several types of future scenario-based methods and providing principles to guide the reader in their use (Wollenberg et al, 2000).

Exploring Biological Diversity, Environment and Local People's Perspectives in Forest Landscapes: Methods for a Multidisciplinary Landscape Assessment – gathering natural resource information that reflects the needs of local communities, based on work with communities in Indonesia and including case studies and methodologies (Sheil et al, 2003).

Beyond Fences: Seeking Social Sustainability in Conservation – the process of social sustainability covering methods for involving people, including a series of participatory tools and many case studies (Borrini-Feyerabend with Buchan, 1997).

Participatory Approach to Natural Resource Management – guide covering participation planning, individual and group methods, public events, instructions for facilitators and more, from a temperate and boreal perspective (Loikkanen et al, 1999).

Tree and Land Tenure: Rapid Appraisal Tools – guidelines for rapid appraisal methods to gather information on tenure and natural resource management; an overview that also suggests further reading (Schoonmaker Freudenberger, 1994).

Community Forestry: Participatory Assessment, Monitoring and Evaluation – a very detailed manual on the options for collaboration with local communities in assessment and monitoring (Davis-Case, 1989).

Rapid Appraisal for Community Forestry: The RA Process and Rapid Diagnostic Tools – detailed guidance on how to carry out a rapid rural appraisal within the context of community forestry including diagnostic tools (Messerschmidt, 1995).

Future Scenarios as an Instrument for Forest Management: Manual for Training Facilitators of Future Scenarios – succinct and useful guide to help train users of scenarios in forest planning (Nemarundwe et al, 2003).

As an addition to the methodologies outlined, some specific notes regarding participation in the context of forest quality assessment are given below.

- **A local facilitating body is an extremely useful part of any participatory assessment**
  When the system is being used in whole or part as a way of negotiating or seeking agreement for particular aims, the body initiating the assessment will be hampered if it is also facilitator. In Wales we worked with a local NGO that already had contacts with all major stakeholders. In Africa, WWF was the facilitating body and in Central America IUCN fulfilled this role. Using a local facilitator has several advantages. It speeds up the process of locating and contacting stakeholders, and contact from a 'known' body with a previous track record in the area is likely to be less threatening than from an outside body. The local body can provide advice about likely responses, local politics, personality clashes and history that an outside body will need a great deal of time to learn, and in addition having a separate facilitator leaves the initiating organization free to take part *as a stakeholder* with opinions and needs rather than having to maintain a neutral stance. A local organization will also in some cases help with language(s).
- **Participation needs to be adapted to particular cultural and geographical situations**
  We started by assuming that participation would usually be in the form of group meetings or workshops where all stakeholders would get the chance to hear various points of view and to express their own thoughts. We quickly found that this was not suitable for all situations. Problems of timing and geography make it difficult for all stakeholders to meet at one time and those that cannot attend a meeting will tend to have less influence on the process. Problems of political influence and power will stop or inhibit some groups. In Guatemala the historical legacy of civil war made it impossible for stakeholders to meet. Different stakeholder groups will not necessarily even share the same language. Rather, the participation should be tailored to the particular needs of each situation and involve a range of approaches including:
  - workshops;
  - one-to-one meetings with individuals;
  - meetings between researchers and particular stakeholder groups;
  - telephone conversations, email contact, contact by letter;
  - use of article in the local press and local radio stations;
  - posters.
  No single exercise can contact all stakeholders. In addition, many will not be interested enough to want to devote time – an attitude of 'you have to participate' is not to be encouraged.
- **Timing of stakeholder involvement needs to be decided on a case-by-case basis**
  A decision about when to involve stakeholders, and which stakeholders to involve, is a key issue relating to a negotiated or participatory assessment. Instigators with a particular interest (such as an NGO with a conservation agenda) may wish to make decisions about when and if to talk with other interest groups on a pragmatic and tactical basis. The earlier stakeholders can be involved the better in terms of building trust. However, where some stakeholders are likely to be antagonistic to the aims of the assessment or associated project, it may be worth building consensus amongst supportive and neutral stakeholders first.
- **One of the key stakeholder roles will be in choosing the type of output**
  Different outputs are needed for different processes. The NGO we worked with in Wales wanted a vision of where their work should go and were most interested in background data, a written report and a set of proposals. WWF in Cameroon and Gabon was interested in scoring the quality of forests. IUCN in Central America want to use the assessment to measure progress on the Central American Biological Corridor. All require very different outputs.

Choosing an approach

The degree and timing of participation will therefore depend on what the assessment is being used for and to some extent on the political judgement of the assessors. There is no hard and fast rule about when and how to involve people, but addressing the following questions may help to make a decision:

- Will local communities be directly affected by decisions made as a result of the assessment?
- Are people likely to be interested?
- Will people participating in assessments get anything useful out of it, or are they just helping you?
- Could any stakeholders (particularly more vulnerable communities) be damaged by a completely open process?

Stakeholders can be involved in a variety of ways, ranging from public meetings to one-to-one discussions as appropriate; techniques such as participatory mapping may also be appropriate. Approaches to participation can also be active or passive – that is, actively seeking out stakeholders or allowing them the opportunity to comment (for instance through inviting written contributions or putting up material on a website). A whole range of different approaches might fit different situations and one aspect of a successful and responsible approach is not to claim too much. In general, except for rare cases where public feelings about an issue run very high, the more effort you put into collecting opinions and information, the more you will get. Sticking a notice on an obscure website is not the same as active engagement with different stakeholder groups and it is important for any assessment to be clear about who has had the opportunity to take part and how that opportunity was presented.

4 How to Assess Forest Quality at a Landscape Scale

We've been walking for several hours along tiny paths, without seeing anyone else. From the ridge of the Jura Mountains, between France and Switzerland, we can look down on a mosaic of forest and farmland stretching to the Lake of Geneva. Although it feels remote up here, these mountains are subject to intensive use. Farmers march their cows up to the summer pastures, adorned with flowers and bells in a popular ceremony stretching back for centuries. The commune manages the forests in a complex system that maintains a continuous cover of trees. And people flock here for fun, to walk, ride mountain bikes, go cross-country skiing and to hunt the chamois. Villages are scattered over the lower slopes and a tiny train chugs between Switzerland and France. Yet much of the land remains essentially natural, with abundant fields of wildflowers, and healthy populations of deer, wild boar and even the shy and elusive lynx. Most of the values of the Jura draw on the whole landscape rather than a random collection of woods and fields.

The landscape is more than the constituent parts. Ralph Waldo Emerson captured the essence of a landscape back in 1836:

> *The charming landscape which I saw this morning, is made up of some 20–30 farms. Miller owns this field, Locke that and Manning the wood beyond. But none of them own the landscape. There is a property in the horizon which no man has but he whose eye can integrate all the parts, that is, the poet. This is the best part of these men's farms, yet to this their land deeds give them no title.*

Forest quality assessment is based around the recognition that a full range of biological and human values from forests can only be reflected – and therefore only be meaningfully assessed – at a landscape scale. Rather than simply looking at an individual site, the assessment method examines how the values of many sites combine within a landscape. Although the assessment was originally developed as a way of finding the status of forest quality in a particular area, implementation has thrown up several different ways in which it can be applied. Indeed, the assessment system described is perhaps better regarded as a framework that can be used in a variety of ways and to achieve several different outcomes.

The following section describes a framework for assessing forest quality. Within this framework there is a lot of room for tailoring the system to individual needs, in terms of the aims of the assessment, types of indicators used, the time and detail of the assessment, the degree to which it is participatory or expert driven, and the analysis and presentation of the results. A six-stage process is presented in Figure 4.1.

Before outlining the framework in detail, we look at *different levels of assessment*.

Levels of application of the forest quality framework

Within the basic framework outlined above, assessment can vary in detail and approach depending on the time and resources available. The design of the assessment should be influenced to a large extent by what it is being used for. At one extreme, an assessment might be used once and for a single purpose, for example in a planning process, a research project or to identify high conservation value forests in the area. Such studies can be as time-consuming and detailed as the needs and budget determine. At the other end of the spectrum, assessments can be used as part

Stage 1: Identifying the **aims** of forest quality assessment

Stage 2: Selection of the **landscape**

Stage 3: Selection of the **toolkit** (indicators)

Stage 4: Collection of **information** about the indicators

Stage 5: **Assessment**

Stage 6: **Presentation** of the results

FIGURE 4.1 **A six-stage framework for forest quality assessment**

of regular monitoring and in this context might be fairly rapid analyses of a few key elements for monitoring. The practical options vary depending on budget and on need, and a summary is given in Figure 4.2.

At its simplest, a forest quality assessment can entail the completion of a basic scorecard to summarize initial impressions about the status and potential of a forest landscape – a few hours' work – while a full assessment to develop a data-rich result could grow into a research project of many years duration. (Although we considered options for assessment by literature survey, experience showed – see Case Study 4 on Switzerland – that there are few, if any, areas where this alone will provide enough detail for a meaningful assessment.) The system should remain flexible enough to meet local needs and particular situations. The framework can be applied at many different levels of detail. Below three different types of assessment are summarized as examples – these are by no means the only options:

- *Scorecard* – a simple, standardized approach to summarizing information, and if desired also 'scoring' quality, suitable for initial planning purposes or a quick and regular monitoring and evaluation of project outcomes, usually completed by one person or a small number of specialists. A draft scorecard is presented in Appendix 1; this can be adapted to local conditions.
- *Rapid stakeholder-driven assessment system* – a fairly quick assessment (2–3 weeks on average) driven largely by literature research, workshops and one-to-one meetings with complementary research, suitable for more detailed planning or as the basis for negotiation of land use at a landscape scale, usually by implication involving local stakeholders although this is not essential. Here 'stakeholders' can include both professionals and local people with specialized knowledge of their own landscape.

One-off assessment

Options for:

- Detailed surveys

- Interviews

Continual monitoring

Options for:

- Minimal surveys

- Use of existing data wherever possible

FIGURE 4.2 **Detail of assessment depending on use**

- *Detailed data-driven assessment system* – a longer-term assessment, with data collection and fieldwork involving a range of specialist disciplines, suitable as part of a wider research project looking beyond immediate planning needs, with or without stakeholder involvement.

Each of these puts a slightly different emphasis on the six-stage framework, as shown in Table 4.1.

TABLE 4.1 **Approaches to forest quality assessment**

Stage of framework	Scorecard	Rapid assessment	Detailed assessment
Identifying aims	Aims are limited: initial survey or monitoring and evaluation	For planning and negotiating landscape scale activities	As part of a longer-term research project
Identifying the landscape	Usually by experts (or predetermined)	Usually by stakeholders	Either by stakeholders or experts
Selecting tools (indicators)	Use of standard indicator list in most cases	Usually by stakeholders	Usually by stakeholders
Collecting data	Reliant on existing knowledge	Mainly through workshops, bilateral meetings and literature	Through workshops, meetings and literature, backed by fieldwork
Assessment	Based on standardized approach	Based on needs of particular assessment	Based on needs of particular assessment
Presentation of results	As either a score or a standardized data sheet	Based on needs of particular assessment	Based on needs of particular assessment

The framework is designed to be flexible and approaches can be combined and modified to meet the needs of individual users, using the various options outlined in Figure 4.3.

In the following sections, each stage of the framework is discussed and some options for development are identified.

People involved – a forest quality assessment can be carried out by:

- experts or specialists working on their own;
- specialists working in association with a few key stakeholders;
- a fully participatory process involving participatory rural appraisal (PRA) or similar approach.

Collection of information – can involve some or all of the following:

- primary field research;
- interviews and workshops;
- use of geographical information systems (GIS) or other remote survey techniques;
- literature surveys and use of existing data.

Analysis and presentation of results – can involve some or all of the following:

- written report;
- maps and diagrams;
- photographs or PowerPoint presentations;
- a scoring system;
- a SWOT (strengths, weaknesses, opportunities and threats) or similar analysis.

Use of results – can also cover several different options:

- planning a landscape approach to conservation;
- negotiating conservation strategies with local communities;
- long-term monitoring of the status of forests in a landscape;
- monitoring progress in implementing conservation projects;
- assessing specific elements of forest quality as part of wider research.

FIGURE 4.3 **Different ways of applying the framework for forest quality assessment**

Stage 1: Identifying the aims of the forest quality assessment – why, where, who and how

The type of assessment is influenced by what it is being used for, where it takes place, who is involved and how the assessment team goes about its work.

Before starting an assessment, four basic questions need to be addressed:

1 *Why* is the assessment being carried out?
2 *Where* will it take place – and particularly what are the boundaries of the assessment area?
3 *Who* will be involved – 'experts', local communities, managers?
4 *How* will assessment be carried out?

This probably sounds obvious. Unfortunately, a lot of assessments take place without these very basic questions being addressed and a lot of time and resources get wasted in consequence.

Terms of reference can be useful, which should include: main aim, participants and location, timeline, planned methods of assessment and planned dissemination of results. Discussing, modifying and agreeing these parameters are themselves early stages in the process of assessment.

WHY IS THE ASSESSMENT BEING CARRIED OUT? MAIN AIMS

There are many reasons for carrying out a forest quality assessment, for instance:

- designing a protected areas system;
- planning extractive reserves;
- improving a suburban recreational area;
- resolving conflicts between conservationists and small forest owners;
- negotiating a concession for timber or non-timber products;
- developing an ecoregional conservation plan;
- developing a monitoring and evaluation system;
- monitoring progress towards forest landscape restoration or sustainable forest management;
- identifying high conservation value forests (Jennings et al, 2003);
- as part of an advocacy process to make a case for conservation or development.

One key distinction is whether the assessment is likely to be a *once-only process* for planning, negotiation and development of a project, or whether it is a *continuing process* monitoring change over time.

When repeated assessments are needed, the importance of finding accurate but realistic indicators is correspondingly greater; most monitoring and evaluation systems fail because data collection proves to be too expensive or time consuming to sustain. Long-term monitoring and evaluation systems have to balance the ambition of producing a thorough picture with the need to keep making measurements. In these circumstances, unless long-term funding is found, most indicators will probably need to draw heavily on data that are already collected. Existing statistics will need to be analyzed and moulded to give useful information; collection of new data should be limited to things that are currently unknown and are also vital to measuring progress. These issues are examined in greater detail in Case Study 5 on Viet Nam. In the case of monitoring and evaluation systems, an

---

| Box 4.1 | **Setting a vision for the landscape** |

In the cases where assessment is being used as part of a planning or negotiation process, the assessors themselves will often need to have an initial idea of their own vision for the forest landscape before starting the assessment, if only to have a clear idea of the key priorities for assessment (which for example will influence choice of indicators). Setting a vision for the landscape could therefore be a key part of this stage in the assessment. These ideas are likely to be modified following the assessment.

important decision is whether it is a *neutral measurement of progress* or a *tactical tool that can help to drive progress*. Getting agreement from other stakeholders on measurement of a particular indicator can be a factor in making sure that the indicator is fulfilled, for example that protected area targets are reached or that non-timber forest products (NTFP) are maintained in managed forests.

WHERE DOES THE ASSESSMENT TAKE PLACE? LOCATION

The location should be determined by the assessment team and if appropriate also by other participants (and the eventual choice may in turn affect the choice of participants). Boundaries can be drawn along biological, geographical, political, economic or social lines. An official or unofficial management unit will often be the basis of assessment, but edge effects are important – it will often be necessary to extend the assessment beyond economic or political boundaries to reflect real human and/or biological values. In a participatory assessment an idea of the likely landscape might be determined by the specialists but tested in a stakeholder consultation, or might be suggested by other stakeholders. If an assessment takes place over a relatively large area that has been defined for conservation purposes, it may be necessary to define one or more 'cultural landscapes' within the 'conservation landscape' to ensure proper participation by local communities (Maginnis et al, 2004). Selection of the landscape is dealt with in more detail in Stage 2 below.

WHO IS INVOLVED? PARTICIPANTS

An early decision is whether an assessment should be an expert-driven or a participatory process (see Chapter 3). Both are valid in different situations. A participatory process is strongly recommended if the assessment is being used for planning or negotiation of land uses at a landscape scale.

The assessment system assumes that in a participatory process as wide a range of stakeholders as possible is represented. A possible list to be involved in a meeting would include:

- a facilitator who is familiar with the forest quality concept;
- a researcher or researchers (biological and social expertise);
- stakeholders including those with legislative and/or management control, local communities and other interested parties such as non-governmental organizations (NGOs), industries and trades unions;
- observers – in a meeting not everyone present necessarily has to be a participant.

Taking part in assessments carries costs for stakeholders in terms of time and effort and it is up to the assessors to make the process as easy as possible, including if necessary offering payment to cover people's time. Participation is best regarded as a service that stakeholders provide to help improve the assessment.

HOW ARE THE DATA COLLECTED?

Data collection depends on the level of assessment. For a scorecard assessment, previous knowledge amongst the person or persons filling out the scorecard is assumed and probably supplies most of the information needed. In a rapid assessment, existing information will have to be used as much as possible and active data collection (for example field research) necessarily minimized, with new data coming from workshops and interviews. Transparency regarding the source is important and contradictory information can be included if it illustrates different perspectives or different possibilities. Gaps in data need to be clearly identified. In a detailed assessment, field survey work is expected. Possible data sources include (not an exhaustive list):

- information from stakeholders: both facts and also opinions;
- group interviews with stakeholders and others: either particular interest groups or meetings with representatives of many different interests;
- maps: for example of forest cover, settlements, roads;
- government surveys: for example population and employment statistics;
- industry information: for example on timber production, number of companies operating, employment;
- historical records: for example on the presence of forests in the past, sites of cultural importance;
- academic papers: for example social trends in the area, studies of indicator species;
- biodiversity surveys: for example on the presence of species dependent on certain forest types.

In many cases, probably most cases, information will be incomplete; the final assessment will need to include reference to both the accuracy and coverage of information.

HOW ARE THE DATA INTERPRETED? ASSESSMENT

The aims, participants and location will all have an influence on the way in which the results are assessed and presented. As in collection, data interpretation can be driven through an expert or participatory process, or some combination of both, and can involve standardized and informal assessment methods. These issues are explored in more detail below.

Stage 2: Selection of the landscape

Deciding on the size and shape of the area to be assessed is itself often an important part of the assessment process.

The science of landscape ecology aims at understanding the environment at an intermediate scale between a detailed study of a specialized system and the general analysis of an entire country or biome. Landscapes – both natural and modified by humans – are generally made up of a mosaic of different types of habitat. Much of the thinking behind this approach rests on the assumption – itself based on experience – that the landscape is the scale at which different land uses can best be balanced.

Common sense suggests that a 'landscape' is necessarily a fairly approximate measure and that the size and the boundaries of a forest landscape will change from one place to another, depending in part on the reasons for an assessment. Because 'landscape' is itself a social construct, different stakeholders will have their own views about the shape and boundaries of their own landscape. A 'conservation landscape', which is often defined in terms of viable populations of large species and can therefore be very large, is likely to have a series of 'cultural landscapes' embedded within it or overlapping with it (Maginnis et al, 2004).

In practical terms, someone has to decide on the area to be assessed, but the complications of interlocking and overlapping landscapes should be noted. Landscapes that might be suitable for a forest quality assessment include, for example: a protected area and its buffer zone (for example planning and assessment within the United Nations Educational, Scientific and Cultural Organization's (UNESCO) biosphere reserves); a timber or mining concession and the surrounding area; indigenous peoples' territory; the range of a particular species in the case of developing a conservation plan for the species or using it to set the scale of a conservation landscape (Sanderson et al, 2002); a legislative area such as a village or county; a watershed; a priority conservation landscape in an ecoregion (Loucks et al, 2004); a small land area such as an island (probably with some surrounding marine habitat and perhaps also with links to neighbouring islands if these are close); or a commercial forest and surrounding farmland and settlements.

In the current context therefore landscapes can vary in size from a few tens of hectares to thousands of square kilometres. Although the same basic approach to assessment can be used for them all, the amount of detail will vary between a relatively small area that can if necessary be surveyed quite thoroughly and a large area where assessors will be reliant on far coarser and more approximate indicators.

Deciding on the boundaries of the landscape should therefore be one of the first stages in the assessment. In some cases this will be fairly obvious and will become clear as soon as the reason for carrying out the assessment is explained. In other situations, the decision will necessarily be more complex. Some initial guidelines are suggested below.

SOME GUIDELINES FOR DEFINING THE LANDSCAPE

Landscape needs to be determined with respect to the following factors:

- Motivation: choice of landscape will be affected by the reasons for undertaking an assessment. For instance, if a company wants to assess the quality of its own forest land, the landscape will be based around this unit (although perhaps spreading beyond to identify possible impacts in surrounding areas). A forest quality assessment initiated by villagers or by conservation biologists will start from different perspectives.
- Boundaries: in some cases the boundary will be clearly and/or legally defined, while in other situations its border may be less precise. Edge effects may be important – for example should an assessment of a protected area also include the buffer zone or the traditional homelands of people who live partly within the protected area? Because everything interacts the temptation will be to keep on expanding the landscape but at some stage a decision needs to be taken about what is left in and out
- Size: will depend on the primary motivation for carrying out an assessment and could for instance be driven by biological factors (for example viable population size or range for particular species) or political factors (for example forest management unit area). Although it is dangerous to generalize, in very varied conditions or areas of high human population or political complexity, the size of the landscape that can realistically be assessed in any detail is probably smaller than in simpler ecosystems or places with smaller populations. There is a trade-off between the elements of complexity, size of area and detail of assessment that needs to be taken into account when planning the assessment.

In the end, landscapes are likely to be defined by reference to one or more of the following factors:

- ecological or biological;
- geographical;
- political;
- economic;
- social.

Some definitions and examples are given in Table 4.2 to illustrate the range of issues that assessors should consider when defining the landscape.

TABLE 4.2 **Some perspectives on 'landscape'**

Determining factor	Example
Biological	Size needed to maintain viable populations of all species Ecoregion Ecosystem
Geographical	Watershed Natural forest boundary Delta Island Mountain range
Political	Legislative area, such as local authority area, state forest Protected area
Economic	Company owned area Concession area Plantation area
Social	Language group Cultural group Group sharing similar religious beliefs Economic group 'As far as the eye can see' 'Community'

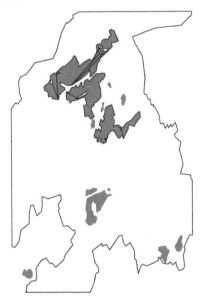

Note: In Switzerland, we tested the approach through an assessment of the forest estate of the city of Solothurn, which exists in a series of different patches, interspersed with other forests, farms and villages. Many of the landscape indicators became impossible to use without including the land 'in between' the patches. Important issues that relate to the forest in the context of the rest of the landscape were lost. Whatever scale or shape of landscape is used, it should exist as a whole and not be broken up into different components or contain significant areas omitted from the assessment. Defining the landscape thus becomes a key part of the process, and is now reflected in the methodology.

FIGURE 4.4 **The system only works for coherent, integral approaches**

SOME GENERAL LESSONS LEARNED ABOUT LANDSCAPE IN FOREST QUALITY ASSESSMENTS

There is a trade-off between an ecologically coherent landscape and data quality

One of the underlying aims of working at a landscape scale is to shift the emphasis of decision-making about forests from political units to those with ecological or social coherence. However, this makes collection of information more difficult if the landscape cuts across political boundaries. When collecting information in Wales, statistics on employment, timber production, population, roads and even biological records from published sources could be used as a general indication of condition, but they had been collected over different landscapes that did not coincide neatly with the political units selected by the local community.

In many cases the advantage of working in a coherent landscape is worth the sacrifice

If only poor quality information is available then this is part of the knowledge gained by the assessment. In more ambitious assessments, this also shows where new research or surveying is needed. Poor data only really become dangerous (or at least becomes much more dangerous) if they are assumed to be accurate. One option to address variable levels of information is to include a rating for data quality and this is addressed in Stage 5 below.

The concept of 'landscape' changes with culture

A persistent criticism of the use of the word 'landscape' is that it has existing meanings. We recognize this but continued using 'landscape' because it conveys the right kind of 'image' to most people. Those who have the greatest difficulty tend to use it already with precision in another context. However, the above relates to the use of 'landscape' in English and for instance use in French and Spanish created some problems despite efforts to find accurate translations. It became clear during field-testing that 'landscape' means different things in different cultures. Populations/communities established for long periods in a particular location tend to have a stronger sense of landscape than more recent settlers. In Wales, where families often trace their history back generations, the concept was easily understood. In Central America, where many communities contain recent migrants and illegal settlers, the concept was more difficult to explain. The consultants finally put a figure on what they meant (something between 25,000 and 150,000 hectares) to get the idea across.

Scale of landscape also changes in different parts of the world

One of the clearest messages to emerge from using the forest quality assessment methodology in four different continents is that the practical scale of the landscape differs in different geographical and political situations. As mentioned above, the landscape that people are capable of working in generally contracts in more crowded or politically complex situations:

- In Wales we worked on an area of around 25,000 hectares.
- In Central America The World Conservation Union (IUCN) worked on 25,000–150,000 hectares.
- In Africa World Wide Fund for Nature (WWF) considered areas up to a million hectares.
- In Viet Nam a monitoring and evaluation system is being developed for eight provinces.

All of these scales made sense in their own context. Where larger areas were assessed it was either because the landscapes were simpler and human population less, or because data needs were simpler.

Note: While the concept of 'landscape' was well understood in the long-settled areas of Wales; it caused far more confusion (and concern) in recently settled areas of Central America, particularly in situations where land tenure remained unsecured.

Source: Nigel Dudley

FIGURE 4.5 **The concept of 'landscape'**

Stage 3: Selection of the toolkit (indicators)

The assessment uses a criteria and indicators (C&I) approach. The indicators used are either chosen on a case-by-case basis or modified from an existing list. They are either chosen by a single person or small group of experts, or selected through a participatory process involving a range of stakeholders.

Indicators are exactly what they say – things chosen to indicate overall status without undertaking a total survey of everything. The great Argentinean writer Jorge Luis Borges wrote a story about a map that became so detailed that it grew to the same size as the country that it was attempting to portray; now abandoned, scraps of it can still be seen blowing around the kingdom (Borges, 1962). Indicators are a way of avoiding the problems that Borges' cartographers encountered.

Because indicators can only ever tell part of the story, there are limitations on their accuracy and the 'wrong' indicators can easily produce a distorted or biased picture of reality. For example, using a species of bird as an indicator of biodiversity only works if that species is likely to be affected by change: if the bird is a resilient and adaptable species then the fact that its population stays stable may mask the fact that other species have declined or disappeared. Alternatively, a well-chosen indicator can, if properly analysed, tell many different stories about an area.

For example, in the middle decades of the 20th century the status of reindeer (*Rangifer tarandus*) and moose (*Alces alces*) were used as a major indicator of forest condition in managed forests in northern Scandinavia. Both these choices were justifiable: the reindeer has enormous cultural and economic importance to the Sami people and the moose is an important source of food to many families in Arctic Lapland. Yet these large mammals can survive in secondary forest where many other species will decline or disappear. Continued loss of old growth forest, and of key forest components such as standing and lying dead timber, took place while the large mammal populations were maintained, resulting in for instance the listing of several hundred saproxylic (dead wood living) species on the IUCN Red List of endangered species for Sweden (Fridman and Walheim, 2000).

In a participatory assessment, the range of indicators used in any given assessment should be chosen by both the assessors and local stakeholders. In many cases assessors might start by suggesting a possible list of indicators that could be used and that stakeholders can add or subtract from: this is a trade-off as it certainly helps people start thinking but also tends to skew the discussion towards what the 'experts' suggest. A preliminary list of indicators is given in Table 4.3.

Due to the wide range of different values that are included, the assessors have an additional problem that many indicators will provide very different *types* of information. Comparison of the spiritual value of a forest with its biological quality for instance poses considerable challenges.

Many other indicators, while providing the type of information needed, require types and quantities of data that will be out of the reach of most assessments. In these cases, assessment may have to take the form of identifying broad categories (such as 'very important', 'important', 'significant', 'fairly unimportant' and 'irrelevant', or scoring on a numerical scale). Although such judgements are approximate, if knowledgeable local stakeholders support them, they can still provide a valuable source of information.

GENERAL PRINCIPLES FOR THE SELECTION OF INDICATORS

Selection of indicators depends on the particular needs and aspirations of the assessor and participants. However, some general principles may be helpful:

- *Range of indicators*: indicators should in all but exceptional circumstances cover the full range of forest quality values, rather than concentrating on a few areas. For example, a survey of biodiversity values that does not also take account of social issues, commercial uses and so on will only provide a very partial picture of what is happening in the area. Many scientists feel uncomfortable about using different levels of data quality, and tend to jump automatically back to the better studied issues where data exist or can easily be collected, but by doing so they risk missing key factors. The extent to which spiritual sites are important in determining land management in many countries is a clear example; because these values cannot be measured in a conventional manner they have tended to be ignored although they have major management implications (Sochaczewski, 1999).
- *Breadth of indicators*: the best indicators can provide information on several different aspects of quality. For example, information on distribution of the age-class of trees says a great deal about the natural ecological patterns and likely functions of the forest, gives information about economic value and also has relevance for many social issues such as recreation and hunting.
- *Partial data sets and quality of information*: almost all data sets will be partial. This does not necessarily matter so long as the limits of the information are clearly acknowledged during and after the assessment (indeed knowledge about lack of data can itself be a valuable outcome of the assessment).
- *Data availability*: the method by which each indicator is measured will depend to some extent on data availability. The presence of good GIS information, biodiversity surveys and data relating to timber and forest management will for instance have an impact on the ways in which indicators are represented. In areas where there are little or no data, either baseline surveys will have to be carried out or the assessment will rely much more heavily on the judgements of stakeholders and consultants.
- *State, pressure and response*: most of the indicators suggested are *state* indicators; that is, they refer to the current status of the forest landscape. However, a few can also be – or in some cases are specifically designed to be – *pressure* and *response* indicators, that is, they show pressures that may change the forest landscape in the future and likely responses. This is examined in

more detail below. State indicators that are measured repeatedly will provide information on trends, although more immediate information on changes can sometimes be gained through analysis of pressure and response indicators.

In the following framework in Table 4.3 a total of 24 indicators are suggested; most of these have a range of different types and units of measurement. Each indicator is explained and some assessment methods suggested in the following section. This is not a complete list of possibilities but a skeleton that can be supplemented or replaced by alternatives as desired.

TABLE 4.3 **A draft list of standard indicators for a forest quality assessment**

Indicators of authenticity	
Composition	Composition of species, ecosystems and genetic variation
Pattern	Spatial variation of trees with respect to age, size etc.
Function	Continuity, proportion and type of dead timber etc.
Process	Disturbance patterns, life cycles
Resilience	Tree health, ecosystem health, ability to tolerate environmental stress
Continuity	Area, degree of fragmentation and age
Development patterns	Management choices and other pressures
Indicators of environmental benefits	
Biodiversity conservation	Of ecosystems, species and genetic variation
Soil and watershed protection	Reducing erosion, maintaining fisheries, controlling flooding
Impacts on other ecosystems	Such as freshwater and coastal zones and through afforestation
Climate stabilization	Transpiration, regulation of climatic extremes, carbon sequestration
Indicators of social and economic benefits	
Wood products	Fuelwood, charcoal, timber, pulp, paper, reconstituted fibres
Non-wood products	Food, oils, medicines, aromatics, resins, dyes, building materials etc.
Employment and subsistence	Jobs in forest management, hunter-gatherers, use of fuelwood
Recreation	Walking, hunting, sports, camping, mushroom gathering, camping etc.
Homeland	For indigenous and local people
Historical values	Including historical artefacts and historical management patterns
Cultural and artistic values	For painters, writers, musicians and as a source of inspiration
Spiritual values	Sacred trees, sacred groves, burial sites, use for spiritual fulfilment
Management and land use	Types of management system, incentives etc.
Rights and legal issues	Access to forests, ownership, traditional rights
Knowledge	Indigenous and traditional knowledge, use for scientific research
Nature of incentives	Political, cultural, economic, social, spiritual etc.
Local distinctiveness	The importance of a particular place to individuals or communities

CHOOSING INDICATORS IN A PARTICIPATORY MANNER

The issue of participation is discussed separately in Chapter 3, however, some specific notes on participatory approaches to choosing indicators are given here.

The selection process: the process of selecting a set of indicators will generally start with the introduction to the concept – preferably through a pictorial or other visual presentation in the case of a larger meeting, perhaps followed by a site visit. The facilitator then has two options for deciding the set of indicators:

1   The facilitator introduces a proposed set of indicators and participants add or reduce from this list;
2   The facilitator works with the group to produce a list of indicators from first principles.

In practice, the difference may be more one of presentation and based on the temperament of the people involved. In the case of one-to-one meetings, simply finding out what is important to people will start to build up a list of indicators. As more is learned about the system, fairly standard toolkits for different situations will probably start to emerge. Where practical, surveys should use indicators that pass tests of relevance, reliability and accuracy. Once the indicators have been chosen, they may also be ranked in terms of their importance. One option is to assemble a matrix as illustrated in Table 4.4.

TABLE 4.4 **Ranking indicators of forest quality**

Ranking	Table of indicators: Criteria of forest quality		
	Group 1: Authenticity	Group 2: Environment benefits	Group 3: Social and economic benefits
High importance			
Medium importance			
Low importance			

THE NUMBER OF INDICATORS

There is no set limit on the number of indicators used in the assessment, although time and resource constraints mean that numbers should be limited to levels that can be measured and assessed effectively. Some indicators may emerge from the data collection process itself. There is little general agreement about the ideal number of indicators. The regional criteria and indicator processes (such as the Ministerial Conference for the Protection of Forests in Europe and Montreal Processes (Canadian Forest Service, undated)) tend to have 60–70 indicators while the Convention on Biological Diversity (CBD) has attempted to set a series of 7–8 key global indicators of biodiversity (UNEP, 1997). Generally, the more indicators that are available, the richer the picture developed, although too much information can conversely confuse the analysis.

DEFINING INDICATORS

It is suggested that indicators might best be defined in relatively broad terms at the stage of setting the toolkit, and researchers encouraged to search for as much detail as possible with the time and resources available. In a rapid assessment the degree of sophistication of indicators will, to a large extent, be related to the current state of knowledge.

TEMPORAL AND PRESSURE INDICATORS

The forest quality assessment method presents a 'snapshot' in time. If the assessment aims also to indicate important trends, some of the indicators should be chosen either because repeat measurements are likely to show changes or because the indicator itself illustrates a trend. Trends will only become apparent if repeated assessments are undertaken over a period. In most cases changes are likely to be subtle so that quite long time periods will be needed to identify definite trends. Including some temporal element in the assessment is therefore important in order to capture as much information as possible.

This issue can be addressed in two ways:

1 By routinely including a pressure, response or trend element in each indicator measured;
2 By including specially selected pressure or response indicators to give a picture of the likely trends.

Some combination of the two is also possible. Many of the indicators discussed above could have a pressure element added on without causing assessors much additional effort. In some cases, particular pressure indicators may also be worth considering, providing specific information beyond that given by an extension to state indicators.

In Tables 4.5 and 4.6 examples are given of how pressure indicators might be added to state indicators to provide another level of information and there are also some examples of indicators tailored to provide information on trends.

TABLE 4.5 **Examples of pressure indicators**

Criteria	Examples of pressure indicators
CRITERION 1: Authenticity	
Natural forest	Presence of illegal loggers
	Presence of mining concessions
	Presence of logging concessions
	Increases in migration into the area
	Changes in forest cover over time
	Increases or decreases in number of Red Listed species
CRITERION 2: Environmental benefits	
Watersheds	Plans for hydroelectric power (HEP) construction
	Presence of large-scale clearcutting
	Current levels of acid deposition
CRITERION 3: Social and economic benefits	
Timber production	Comparison of extraction with maximum sustainable yield data
	Trends in timber demand in key markets
Homeland	Rate of population growth
	Rate of migration to the area
	Changes in socio-economic status of groups living in the area

SOME GENERAL LESSONS LEARNED REGARDING INDICATORS

In addition to the general guidance given above, some more general lessons have become obvious during the project.

Both qualitative and quantitative indicators are needed

Many classically trained scientists are concerned by the frequent use of qualitative data in assessments. However, sticking only to things that can be measured precisely means that many wider aspects of forest quality are left out. Yet some of these are of extreme importance. For example, spiritual values of forests are often impossible to quantify in anything but general terms but can be the dominant factor in determining forest management. 'Quantifiable' may itself often be misleading in practical terms where information is scarce. In the Congo Basin huge areas of the forest are un-surveyed even with respect to mammals such as elephants, and much of their flora and invertebrates are unknown. Inaccurate 'quantifiable' data are often less useful than good qualitative information. The option of trying to fit all indicators into a common format is simplistic and likely to mislead. This does not mean that qualitative indicators cannot be ranked; significance or importance ranking is possible and can help in comparing indicators of very different issues.

Involving stakeholders in indicator selection is useful in a participatory assessment

When a forest quality assessment is being used as a negotiating tool, involving stakeholders in selection of the landscape and the indicators can be a valuable part of the negotiation process. Once all the relevant interest groups have agreed on the indicators then they are in effect also agreeing that the indicators are important and therefore need space in the landscape. If a forest manager agrees that biodiversity is a valid indicator and a conservationist agrees that timber is a valid indicator – particularly if this is done publicly – then the basis for negotiation is already underway.

Public selection of indicators tends to miss those of 'global' significance and to increase the number of social indicators

Experience of selecting indicators revealed that while stakeholders will select a fair range of indicators, these will tend to be directly concerned with the locality – jobs, timber, wildlife, amenity – and will frequently miss issues of global significance such as carbon sequestration and maintaining biodiversity values. Public selection of indicators will also increase the proportion of social indicators. Professional conservationists will generally select a few key indicators, as will the timber trade, but members of the public will normally tend to throw up more detailed issues that directly concern themselves, leading to a far longer social indicator list than that relating to ecology. This does not matter if the assessment is being presented in written or map format but can cause problems if the results are being scored in some way. In Wales, the first public meeting agreed 9 indicators of authenticity, 2 indicators of environmental benefits and 26 indicators of social and economic benefits.

A core set of indicators will probably be required

At the initial workshop to discuss assessment in Switzerland, some of the participants argued strongly against having a core set of indicators, both because of the difficulty in generalizing for every situation and because to do so immediately reduces stakeholders' role in a participatory approach. We started off following this advice but gradually adopted a more relaxed approach that ranged from a completely open-ended system with stakeholders determining forest quality indicators to a fixed list, for instance (continued bottom of p40):

TABLE 4.6 **Examples of trends indicators**

CRITERION 1: Authenticity

Composition	Number of threatened species Number of Red Listed species Number of species that have been extirpated over the past 20 years
Pattern	Changes to the forest mosaic. Comparison with old photographs, evidence of past forest management
Function	Changes in forest cover
Process	Changes in disturbance regime over the past 20 years. Number of fires, major storms, fellings etc.
Robustness and resilience	Threats to tree health Scale and rates of mortality over time Changes in levels of pollutants Frequency of pest and disease attack and trends over time
Continuity	Historical records including maps Tree ring analysis Analysis of pollen records Presence of protected corridors and ecological stepping stones Protected area networks
Development patterns	Plans for future transport and infrastructure development Changes in infrastructure over time (through comparison with old maps etc.)

CRITERION 2: Environmental benefits

Biodiversity	Red List species Other threatened species Species lost from the area
Protected areas	Status of protected area and evidence of management effectiveness
Watershed protection	Soil loss and changes in water quality over time Changes of frequency in flooding Changes in fish stocks

CRITERION 3: Social and economic benefits

Wood products	Changes in volume produced/year and in monetary value
Non-timber forest products (NTFPs)	Changes in collection activity over time Evidence of scarcity or decline in important NTFPs
Employment	Changes in employment over time Future threats to jobs (increasing mechanisation, end of concessions, proposed mill closures etc.)

TABLE 4.6 **Examples of trends indicators** *(continued)*

Recreational value	Changes in number of visitors over time Plans for future tourist attractions/facilities
Homeland	Changes in numbers of people living in the forest Status of land rights claims Evidence for migration into or out of the region
Historical value	Threats to and security of historical artefacts
Aesthetic value	Loss of forest elements with aesthetic value (natural forest, specific features) Likely threats to elements of the forest with aesthetic value Pending designations of Areas of Outstanding Natural Beauty or similar
Educational value	Change in number of educational visits Trends in educational funding, teaching syllabus, infrastructure etc.
Spiritual value	Threats to spiritual sites
Local values	Evidence of changing attitudes
Management and land use	Trends in management, for example: • existence of forest management plans • number and area of certified forests • number, area and type of protected areas
Rights and legal issues	Changing legal structure Challenges to land tenure Evidence of biodiversity prospecting in the area Changes in access rights
Nature of incentives	Changes in types of incentives, for example: • social • political • economic • cultural

- In Switzerland a core set of indicators was chosen by the assessor after consulting the local forestry officer;
- In Wales indicators were chosen in an open public meeting by interested stakeholders;
- In Costa Rica indicators were chosen at an experts' workshop involving several governmental and non-governmental institutions;
- In Cameroon and Gabon indicators were chosen by a local consultant after discussion with stakeholders on an individual basis;
- In Viet Nam indicators were chosen by a research team after over 50 meetings with local and national government and NGO bodies and were then tested in a stakeholders' workshop.

Although involvement of stakeholders is important if the assessment is going to be used for planning or negotiation, for at least some of the purposes of assessment, it will be necessary to

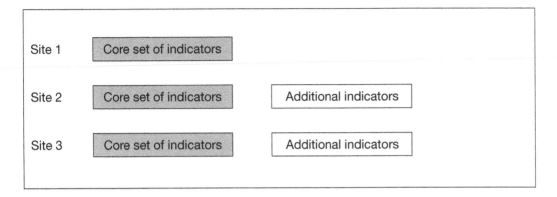

FIGURE 4.6 **A core set of indicators for comparing between sites plus additional site-specific indicators as required**

compare between sites and to compare the status of one site over time. This means having a 'core' set of indicators that is always measured in a forest quality assessment: in this case landscape-specific indicators added by stakeholders during the assessment would be additional to the core set, as illustrated in Figure 4.6.

A core set of indicators is probably essential in cases where the assessment is being used to 'score' forest quality.

Stage 4: Collection of information about the indicators

Once indicators have been chosen and agreed, researchers work (with stakeholders in a participatory approach) to collect the required information.

The type of data required is likely to be influenced by the indicators that are chosen, although some general principles are applicable:

- *Levels of information*: information will be available at international national, regional, local and individual level;
- *Levels of participation*: some data can be assembled almost entirely by the researcher(s) while other information will need active participation by stakeholder representatives and others;
- *Levels of certainty*: quality of information will inevitably vary and it will be up to the data collectors to advise about this and the assessors to respond accordingly, at least to the extent possible.

Information about some indicators may be difficult and/or expensive to obtain and compromises, particularly regarding quality of data and levels of certainty, will have to be made on a case-by-case basis. In cases where a consultant is being used to collect information, a detailed indicator list might provide help for the data collector as outlined in Table 4.7.

APPROACH

Data come from a wide variety of oral, written, statistical and pictorial sources and directly from field observations and studies. Choices about which particular sources to use will depend on time,

TABLE 4.7 **Matrix giving terms of reference for the data collector, with examples**

Indicators	Description	Data sources	Expertise needed
Authenticity of composition	At least species of trees needed, if possible also of other flowering plants and of animals	Published species lists, field surveys	Basic botanical knowledge to understand lists; if necessary level one field assessment experience
Wood products	Ideally volumes of timber extracted and maps of main timber operations (recent past, present and projected)	Information from state forest companies, main timber companies, possibly NGOs	No particular expertise

money and the level of accuracy needed. In particular, there will often need to be a strategic decision about whether the best way to gather information is to pull together 'experts' – who may be local stakeholders or external specialists – or to gather data directly from the field. Comparison of data-driven and workshop approaches with respect to conservation planning seems to suggest that they come up with much the same results (Cowling et al, 2003); a combination of both is also sometimes possible.

DATA SOURCES

Amongst the possible sources of information are:

- *Published information, reports, maps, photographs and so on*: choosing a landscape that may not accord with political boundaries (for example administrative areas) may reduce the usefulness of much statistical information because it can be difficult to extract the relevant portion.
- *Expert workshops*: particularly important for specific indicators or groups of indicators, such as biodiversity, economic values and so on. There are inevitably inaccuracies associated with such workshops for instance scientists tend to know more about biodiversity in the places they have studied than in other parts of the landscape. But at least in a workshop setting there is an opportunity for instant peer review and discussion of various pieces of information.
- *Stakeholder workshops and participatory assessment methods*: for many issues (including those traditionally assigned to 'experts', such as biodiversity) local knowledge may be more complete and more accurate than that of highly trained outsiders. Many methodologies and a great deal of experience have been gained in how best to utilize indigenous knowledge in assessment and planning (Danielsen et al, 2000; Sheil et al, 2003).
- *Interviews with individuals or small groups*: workshops are of limited use in places where for example there are major differences in power relations, languages or perspectives. In these cases stakeholder interviews with individuals or groups may be more useful than working only through workshops.
- *Field research and surveys*: in any but the most superficial studies, some fieldwork will be necessary both to ground-truth information from other sources and to collect data directly on indicators.
- *GIS and other remote sensing information sources*: growing expertise on the use of remote sensing increases the ability to collect information on large and newly defined landscapes, for example remote sensing approaches can now help define forest condition as well as forest cover in many cases.

- *Use of simple self-assessment systems including scorecards*: field research can be simplified through the use of simple scorecards and recording cards, as described for authenticity and more generally below.

Choice of the method(s) will be determined by what the information is needed for, constraints of time and money and whether the assessment is a one-off exercise or the start of long-term monitoring.

MOVING FROM SITE TO LANDSCAPE – SPATIAL ANALYSIS OF INDICATORS

The significance of many of the indicators, particularly but not only those relating to authenticity, depends as much on their location and distribution within the landscape as it does on their total quantity or 'score'. The implication of factors such as timber removals, human population or collection of NTFPs varies widely depending on where they take place. However, recording this in a survey adds additional time and resource costs to any assessment. In practice, the extent to which indicators are spatially analyzed will depend on both the needs of the assessment and the resources available. Several possible options exist:

- *Amalgamation of site studies to cover the whole landscape*: this is probably the most accurate method but is also time-consuming and expensive. However, in many places site-level studies will already be available to draw upon and experience shows that these can frequently provide information of use in a landscape-level analysis as well.
- *Mapping of data*: where possible, presenting information on maps or GIS systems addresses many of the landscape issues as simply as possible and in a form that is likely to be compatible with use for planning purposes.
- *Generalized indicators of distribution*: one option in a quick assessment is to develop some standard format for indicating the distribution of the indicator over space, to be appended to any assessment of the indicator. An example is given in Table 4.8.

TABLE 4.8 **Format for indicating the distribution of the indicator over space**

Indicator	Indicator concentrated in one or two locations	Indicator occurring sporadically throughout the landscape	Indicator spread evenly throughout the landscape
Indicator 1	✓		
Indicator 2		✓	
Indicator 3			✓
Indicator 4			✓

- *Specific categorizations for indicators or groups of indicators*: such as typologies for particular indicators. A typology of authenticity has been developed as one possible way of quickly summarizing information about authenticity (one of several described in this book) and can be found below.

A mixture of approaches is possible. For example, an assessment to plan conservation interventions will want general information about forests at any degree of authenticity but may well want a specific focus on those sites with the highest authenticity (such as High Conservation Value Forests).

Stage 5: Assessment

Data on the various indicators only become truly useful when they are analyzed and the implications worked out.

As with other parts of the process, analysis of data on each of the indicators can either be expert-driven or carried out by a group of stakeholders. Depending on the way in which the assessment is going to be used, this can include:

- a descriptive assessment;
- a scoring system;
- a mapping system using GIS;
- a SWOT analysis.

A combination of all of these approaches is also possible. Each is described in more detail below.

DESCRIPTIVE ASSESSMENT

A descriptive assessment of each indicator and its implications, including maps, charts and so on (see Stage 6) provides the most data-rich source of information, but provides no standardized system for analyzing results or comparing between different indicators or sites. For many purposes, this will be enough.

SCORING SYSTEMS

A score can be useful for comparison between landscapes or as a quick reference for 'strengths' and 'weaknesses' amongst indicators. It can also be useful for presentation and planning. However, if the assessment is perceived as a competition, astute stakeholders could distort the toolkit to raise 'their' score, for example by increasing indicators that show positive quality aspects. Whether results are scored depends on the particular needs of those seeking the assessment. If scoring is used it must not detract from the primary aims of assessment and planning. It is for instance more useful to know that there is disagreement amongst stakeholders than to force results into a neat score. If such a comparison is considered important, we suggest the following options for scoring:

- A simple 'pass or fail' score for each indicator, decided by the experts and/or other participants on a consensus basis. 'Pass' or 'fail' could be judged in a number of ways – for example, whether the quality described by the indicator is sufficiently represented within the landscape, or secure, or even if stakeholders all agree on what actions are needed regarding the indicator.
- Scoring of standardized set of key indicators, measuring:
  - performance of indicator = standardized scoring system;
  - quality of data = levels of certainty.

Performance can be scored as follows:

0 = failed to meet qualities illustrated by indicator
1 = very poor
2 = low
3 = average
4 = good
5 = excellent

An explanation for each of these scores needs to be provided for each indicator.

Quality of data can be scored according to four categories:

Impossible to make a judgement	
Expert judgement only	
Some data, but either limited or dated	
Good data available	

Results are represented on a standard bar chart as shown in Figure 4.7. The main bars indicate scores and the colours below indicate quality of data.

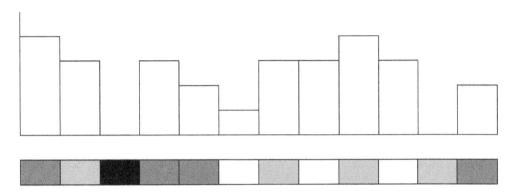

FIGURE 4.7 **Bar chart for scoring forest quality**

A MAPPING SYSTEM

Maps are amongst the most useful ways of presenting information for analysis and are becoming increasingly easy to generate. Mapping data on particular indicators will help analysis by allowing direct spatial comparison of information.

STRENGTHS, WEAKNESSES, OPPORTUNITIES AND THREATS ANALYSIS

A SWOT analysis is where all indicators are listed and described (or scored) as appropriate against four different attributes, as shown in Table 4.9. SWOT analysis is described in Case Study 1 on Wales.

TABLE 4.9 **Matrix for filling out SWOT analysis**

Indicator	Strengths	Weaknesses	Opportunities	Threats

This is particularly useful if the assessment is contributing to a planning process to identify management priorities.

Stage 6: Presentation of results

The way in which both data and assessments are presented is important, particularly if results are to be communicated to a wider audience. Perceptions about what is needed may change during the assessment. For example, if one major gap or problem emerges, stakeholders may prefer to address this immediately rather than spend time writing up the assessment in detail. Alternatively, stakeholders initially intending to carry out a limited, internal assessment may become enthused enough to want to present the information more generally. Facilitators and specialists can advise but should not drive this process.

However, there is sometimes a danger of assessment methods being driven by presentation. We should not sacrifice subtlety of information on the altar of presentational elegance. Many options exist, for example:

- a written document;
- a bar chart showing some form of score;
- maps showing either edited data or scores;
- a management plan;
- a PowerPoint, slide presentation or video;
- a verbal presentation to a wider group of stakeholders.

Examples of some of these are given in the figures below.

FIGURE 4.8 **Written report**

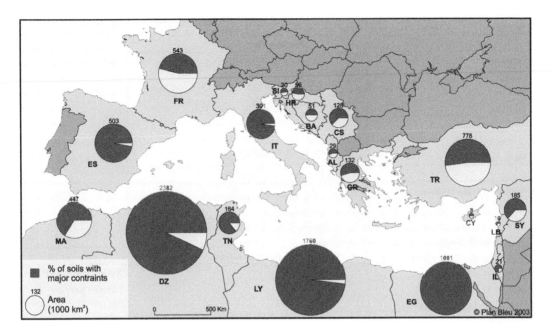

Note: Maps with data included can provide a rapid summary of information and also help put indicators into the context of the whole landscape.

FIGURE 4.9 **A map showing data or scores**

MEETING

Whatever presentational method is chosen, results eventually have to be communicated to a wider group of stakeholders.

SHARING AND LEARNING FROM THE RESULTS

The process does not end with the assessment. A forest quality assessment is only worth carrying out if it is both effectively communicated and acted upon. Forest quality assessments are not simply academic exercises. Further work is required in linking the results to local, regional and national policy. In general, we need to ensure that this remains a bottom-up process and does not inadvertently become a facilitator-dominated process. Further work is also required on how to combine indicators where time or information is limited.

Conclusions: What can a forest quality assessment tell us?

To recap, forest quality assessments can be used for several distinct tasks:

- *Assessing forest condition and potential as part of a planning process*: using a forest quality assessment gives information about the full range of issues likely to be important in a forested

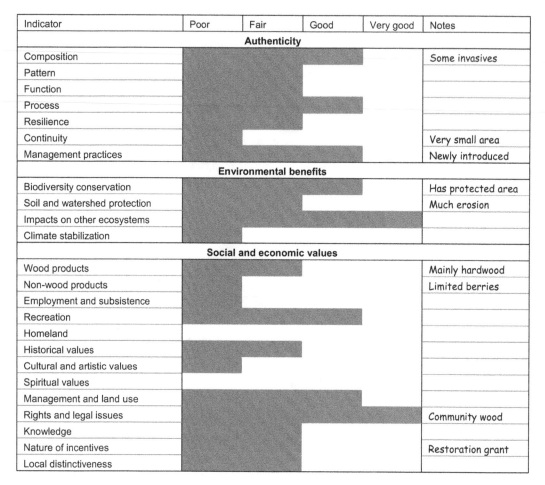

Indicator	Poor	Fair	Good	Very good	Notes
Authenticity					
Composition					Some invasives
Pattern					
Function					
Process					
Resilience					
Continuity					Very small area
Management practices					Newly introduced
Environmental benefits					
Biodiversity conservation					Has protected area
Soil and watershed protection					Much erosion
Impacts on other ecosystems					
Climate stabilization					
Social and economic values					
Wood products					Mainly hardwood
Non-wood products					Limited berries
Employment and subsistence					
Recreation					
Homeland					
Historical values					
Cultural and artistic values					
Spiritual values					
Management and land use					
Rights and legal issues					Community wood
Knowledge					
Nature of incentives					Restoration grant
Local distinctiveness					

FIGURE 4.10 **An example of a bar chart showing scores for different aspects of forest quality**

landscape and thus provides a basis for planning. For example, this use was applied in Wales to draw up an agreed vision for forests within one catchment (see Case Study 1) and could also be used in identifying High Conservation Value Forests (HCVF).

- *Providing the basis for negotiation of trade-offs between interest groups within a landscape*: particularly when a forest quality assessment is used in a participatory process, the selection of indicators and assessment of values can serve as the start of a negotiation process, such as within the landscape approach. For example, this use is being tested in Sichuan, China in association with the WWF China Programme Office.
- *Designing a monitoring and evaluation framework*: the assessment framework can also be used as the basis for developing monitoring and evaluation approaches for projects or programmes. For example, the forest quality assessment was the starting point for development of a draft framework for monitoring forest landscape restoration (see Appendix 2) and this is being tested in the Bulgarian stretch of the Danube River.
- *Identifying priority areas for different interventions*: a forest quality assessment should provide the basis for identification of the most important forest areas from the perspective of both

Note: Stakeholder meetings can either take place in one large event or, more usually, in scattered meetings with different stakeholder groups. Whatever the format chosen, the aim is to provide a forum where people will feel ready to speak out and make their views known.

Source: Sue Stolton

FIGURE 4.11 **A stakeholder meeting**

ecological and social values, such as an identification and mapping of areas of the forest mosaic that are of highest authenticity or dedicated to timber production. For example, this approach was used in assessments in Gabon and Cameroon
- *Contributing to a regional monitoring and evaluation plan*: as conservation efforts are scaled up to include broader regions, monitoring progress on these long and ambitious exercises is both important and increasingly complex. The approach can be used to help to develop such monitoring systems. For example, the methodology was refined, and training materials produced, so that landscape-scale analysis could be included in the regional criteria and indicator process in Central America.
- *Adaptive management*: information from the assessments should also help to modify management interventions over time and thus feed into practical management decisions. For example, monitoring is being used to help adapt management within the Central Annamite Ecoregional Initiative in Viet Nam and eventually in Laos.

It will be clear from all the above that assessment is not a fixed process but needs to be developed and moulded to particular conditions and needs. Decisions need to be made at each stage of the assessment process about what it is trying to achieve, how much detail is needed, who should be involved and so on. The main steps are outlined in Figure 4.12.

Of all the steps, deciding on the way to measure the various indicators is almost certainly the most complex, and accordingly the second part of the book looks at each of a standard set of indicators in greater detail, discussing ways of measurement and sources of information. By way of preparation, an overview of the options for data collection for these indicators is given in Table 4.10.

Stage 1: Identifying the aims of forest quality assessment

What is the assessment trying to achieve?
Leading to related questions:

What level of detail is needed? (scorecard, rapid assessment, detailed research project...)
Who should be involved? (fully participatory, expert driven, key stakeholders...)
Is the assessment a once-off or will it be repeated for monitoring?
Do you need to know just status or also trends?

↓

Stage 2: Selection of the landscape

Is the landscape determined by conservation needs, commercial needs, political divisions, geography etc.?
Who decides?
What are the implications for the assessment?

↓

Stage 3: Selection of the toolkit (indicators)

What indicators are needed? (standard set, chosen by all stakeholders, chosen by experts...)
What is needed to find information on all the indicators?

↓

Stage 4: Collection of information about the indicators

Who collects information? (all stakeholders, consultant, research team...)
How do they go about it? (literature research, field measurements, interviews, workshops)

↓

Stage 5: Assessment

How are the resulting data assessed? (with a score, written assessment...)
Who takes part in the assessment? (by a small group, by all stakeholders...)

↓

Stage 6: Presentation of the results

How are the results discussed and circulated? (report, maps, articles, meetings...)

FIGURE 4.12 **Summary of questions to be addressed in developing a forest quality assessment**

TABLE 4.10 **A draft list of standard indicators for a forest quality assessment**

Key steps in data collection for a standard set of indicators	
Indicator	Data sources
Indicators of authenticity	
Composition	Wherever possible, existing biological records, failing that identification and collection of information on key indicator species to build a picture of overall composition (both native and invasive species)
Pattern	Study of maps, research studies and aerial photographs, also field research looking at for example age structure, canopy pattern etc.
Functioning	Usually best studied through identification and measurement of a few key indicators – species, microhabitats (for example dead wood)
Process	Can be inferred from some of the above, particular age structure of forest, but also from literature, discussion with local foresters and ecologists and direct field observation
Resilience	Health of trees and pressures – in some cases information will exist, if not field research will be important
Continuity	Usually from maps and satellite images (including GoogleEarth) for rapid assessments; also field inspection for more detailed studies
Development patterns	A listing and analysis of key agents of change, probably using one of a series of standard methodologies
Indicators of environmental benefits	
Biodiversity conservation	Existence of official and unofficial protected areas, threatened species (from local or international Red List and from conservation organizations) and details of economically and socially important plant and animal species
Soil and watershed protection	Use of watershed for drinking water, irrigation etc., ideally with economic benefits if these have been calculated
Impacts on other ecosystems	Understanding of likely impacts can come from studying maps and searching literature, in some cases actual field measurements may be needed (for example pollution levels from forest management) or interviews
Climate stabilization	Addressed most directly by an estimate of carbon sequestration (methodologies exist) but ideally should also include ameliorating effects of forest cover for example against drought, sea level rise etc.
Indicators of social and economic benefits	
Wood products	Data from governments, companies and from local communities
Non-wood products	Sometimes government data, more likely interviews, often usefully approached in a workshop

TABLE 4.10 **A draft list of standard indicators for a forest quality assessment** *(continued)*

Employment and subsistence	Statistics but also interviews with companies and with local communities, hard to get precise figures but a general indication is usually enough to make some judgements about importance
Recreation	Government and company statistics, interviews
Homeland	Key need is identification of all groups living in or using the landscape and their relationship with the forest – interviews with communities, local government officials
Historical values	Maps, written records, government records (local museum for example) and interviews
Cultural and artistic values	Written material, maps and interviews
Spiritual values	Interviews with religious leaders, local communities
Management and land use	Use of existing assessments if they exist – for example certification, International Organization for Standardization (ISO) standards, codes of practice – and if necessary also quick assessments of management quality (mainly through interviews with both companies and other stakeholders)
Rights and legal issues	Specific issues of rights (tenure etc.) within the landscape – mainly from interviews and written material if it exists – but also an overview of the relevant legal context in the country or region
Knowledge	A general assessment of the level of knowledge – which should become clear after completing the assessment – important to identify knowledge gaps and different degrees of data quality
Nature of incentives	Understanding of incentives structure, which will mainly be a national-level analysis but can also sometimes include specific local incentives (grants, particular opportunities...)
Local distinctiveness	Interviews, local written sources (local newspaper, local history societies...)

Part 2
Criteria of Forest Quality

This part of the book looks at the three main groups of criteria in greater detail. It discusses the theoretical background to each of the indicators, looks at how they might be measured and provides some sources of further information.

5 Forest Authenticity and Prioritizing Conservation

In this part of Tierra del Fuego, way down at the southern tip of Latin America, the forests have apparently never been cut. Ancient trees, dripping with ferns and lichens, exist in a dense mass of the living and the dead, with huge trunks down everywhere, which make walking a slow process. Groups of black and red Magellanic woodpeckers chatter off noisily as we approach and we can see the final spur of the Andes rising up above our heads. We eat our sandwiches by a beautiful pool created when beavers dammed a small stream. Which is where the idyll of untouched nature starts to unravel, because beavers shouldn't be living within thousands of miles of here: they were introduced by an Argentinean businessman at the end of the 18th century and have proved almost impossible to eradicate, taking up residence even in the Tierra del Fuego National Park. What is apparently a fully natural forest has actually been profoundly altered by one man's action over a century ago.

The first of the main criteria groups provides what will hopefully be a fairly clear series of steps for summarizing information about the ecological value of forests. The issue of authenticity, as outlined below, is the criterion that relates most closely to the values of conservation and least to other direct social or economic human benefits.

Over the past two decades, there have been numerous attempts to define what is important regarding the ecology of natural forest systems by means of defining 'naturalness' – approximately the degree to which the forest corresponds to that expected without any management or disturbance from humans. Terms such as *integrity* (Angermeier and Karr, 1994; Karr, 1994), *naturalness* (Peterken, 1988; Anderson, 1991) and *habitat heterogeneity* (Freemark and Merrimen, 1986) have provided pointers towards ecological value. Some of the most important are summarized in Table 5.1 below (from Dudley, 2003). While these all have their strengths and uses, they are most suitable for application in relatively undisturbed forest habitats.

At present, ecological richness is usually measured in terms of *biodiversity* (for example Hawksworth, 1996). However, it is difficult to establish a scale of measurement for 'natural' levels of biodiversity outside pristine natural forests. Use of species numbers can be misleading; old-growth forests may support fewer species than younger forests but the former often support specialized species unable to live elsewhere. Disturbing forests can create a sudden surge in species numbers, but these are likely to include many weeds and aliens. Species also have different 'values' in terms of their contributions to ecosystem function. Loss or even change in numbers of top predators or herbivores will be more significant than loss of a single invertebrate or lower plant species (Angermeier, 1994).

Another common measure of ecological importance is *age*. The term 'old-growth' has gained widespread usage because of the debate about forests in western North America and in Europe but it has limitations as a general definition, only really being useful in forests where catastrophic change (fire, hurricane) is rare. The future of old-growth forests has become a controversial issue in parts of the world and there have been attempts to define the term with some precision for certain ecosystems (Johnson et al, 1991). However, using age alone as a definition of importance is simplistic. Natural ages of forests differ with habitat, tree species, disturbance regimes and other factors, while in some circumstances 'young' forests may be important from an ecological perspective. Average or maximum age of trees is not necessarily a useful measure.

Authenticity, as used here, is a reflection of the health or resilience of an existing forest in terms of composition and ecology. It often reflects how closely a secondary, managed or disturbed forest resembles the natural forest that it has replaced, but authenticity is more concerned with present

TABLE 5.1 **Some definitions of naturalness**

Definition	Explanation
Ancient woodland	Woodland that has been in existence for many centuries: precise time varies but in the UK 400 years is commonly used (Kirby, 1992b)
Frontier forest	'Relatively undisturbed and big enough to maintain all their biodiversity, including viable populations of the wide-ranging species associated with each forest type' (Bryant et al, 1997). Criteria include: primarily forested, natural structure, composition and heterogeneity, dominated by indigenous tree species. Tends to focus on large and relatively undisturbed areas (for example tundra forests of northern Canada and Russia or tropical moist forests in the Amazon or Congo Basin) and to omit many smaller, fragmented forest types (Bryant et al, 1997)
Native forests	Meaning is variable: often forests consisting of species originally found in the area – may be young or old, established or naturally occurring, although in Australia often used as if it were primary woodland (Clark, 1992)
Old-growth forest	True old-growth: 'stands in which the relic trees have died and which consists entirely of trees which grew from beneath' (Oliver and Larson, 1990)
Old-growth in the Pacific Northwest USA	'A forest stand usually at least 180–220 years old with moderate to high canopy cover; a multi-layered multi-species canopy dominated by large over-storey trees...' (Johnson et al, 1991)
Primary woodland	'Land that has been wooded continuously since the original-natural woodlands were fragmented. The character of the woodland varies according to how it has been treated' (Peterken, 2002)
Wildwood	'Wholly natural woodland unaffected by Neolithic or later civilisation' (Rackham, 1996)

forest condition than about how closely it resembles a theoretical 'original' forest. Although the two concepts will be related, they are unlikely to be identical.

A definition of an authentic forest is one in which *all the expected ecosystem functions can continue to operate indefinitely.*

It is therefore separating the issue of ecosystem health from naturalness (see also Bertello, 1998), although both are important. 'Authenticity', like the other terms referred to above, is an approximation. In this instance, it provides some additional clarification with respect to making judgements about quality from a conservation perspective. Using authenticity helps avoid confusion between total species numbers and species of particular ecological importance. It means that systems with naturally low biodiversity will not be undervalued. It assumes that there is some ranking of ecological importance of species in terms of their role in the overall ecosystem. In general, 'importance' from an ecological perspective will tend to increase with higher trophic levels (that is, a higher place on the food chain), the degree of endemism and the narrowness/fragility of ecological niche, although there are many exceptions to this general rule. Authenticity stresses the importance of natural cycles by for example recognizing the proportion of tree stands of different ages likely to be present in any natural forest area. Authenticity can be used as a baseline against which to measure the conservation and environmental value of disturbed forests. By including composition and process, the authenticity concept can provide guidance on management, forest landscape restoration and conservation strategies. Because it includes overall ecosystem functioning in its definition, it follows that authenticity can best be measured and described at a landscape scale, that is, beyond an individual forest stand.

In a human-dominated planet, authenticity is increasingly compromised. Many ecosystems popularly believed to be 'natural' are intensively and continually managed; some examples are given in Box 5.1.

Box 5.1   **Natural landscapes?**

Although research suggests that up to half the world's land surface may remain with some degree of natural ecology (Mittermeier et al, 2003), many ecosystems are continuing to shrink very rapidly (Hoekstra et al, 2005) and even places that resemble 'wilderness' have often been subtly but profoundly altered, for example through deliberate fire-setting, grazing by domestic livestock, pollution or climate change. For instance, most of the African savannah systems, with their associated high levels of wildlife, are dramatically affected by fire (for example the Serengeti – see Norton-Griffiths, 1979) and vegetation patterns in many savannah protected areas are maintained through deliberate burning, although visitors assume that they are viewing a 'natural' ecosystem. In a by now well-known case, villagers in Guinea, who had long been accused of deforestation, were actually found to be planting trees regularly, maintaining forested habitat in areas that would under natural conditions probably have remained treeless (Fairhead and Leach, 1996). The central Australian desert ecosystem is believed to have been profoundly affected by aboriginal management systems (Flannery, 1994) as is New Zealand (Park, 1995). In Europe, many forests emerged from the last ice age during a period when humans were already actively managing the land, so the idea of a 'virgin' untouched forest is a myth, albeit a persistent and powerful one. Most European countries have less than 1 per cent of their forests in a near-natural state (Dudley and Stolton, 2004). Even in countries where the wilderness ethic is strongest, such as in North America, the landscape has been profoundly influenced by humans (see for example Conzen, 1990), and for instance the sequoia desert ecosystem of southern USA, which for many epitomizes the wild frontier, has only evolved since humans were settled in the area (Phillips and Wentworth Comus, 2000). The concept of 'natural' is highly compromised and needs to be fundamentally re-evaluated.

Basing definitions of importance solely on original state is therefore dangerous from the perspective of biodiversity conservation because many of the most endangered species exist in fragmented and altered forest habitats (indeed this is often the reason why they are endangered). Authenticity would be an unusable concept if it depended entirely on the possibility of reconstructing what will often be an unrecoverable historical ecosystem.

Nonetheless, some idea of natural ecosystem function is important in helping to define authenticity and in seeking to restore aspects of natural forests, for example through forest landscape restoration. The precision with which the baseline can be defined will vary. In some areas natural forest has completely been destroyed; in almost all cases forests have been altered by disturbance, introduction of alien species, pollution, changes to hydrology. Information on which to base notions of authenticity will therefore often be patchy and approximate.

Forest ecologists have to rely on a variety of methods to determine authenticity:

- ecological studies in surviving natural or semi-natural forests – *reference forests*;
- where no natural forests remain, drawing instead on studies of similar natural or near-natural forests found in different parts of the world (particularly useful in understanding 'functioning' and 'process');

- use of historical data, pollen analysis, geomorphological and biogeographical research to determine past vegetation patterns based around historical data;
- predictions of likely vegetation made on the basis of expected vegetation patterns based around geology and geography – for example enduring natural features as used in gap analysis of forest types by WWF Canada (Iacobelli et al, 1994);
- application of ecological theory to fill in remaining gaps in knowledge.

Results will be approximate rather than definitive; but then forests themselves are flexible and adaptive ecosystems. There is a limit beyond which fine-tuning the description will have little practical value. It may be more useful to know whether the forest is becoming *more, or less, authentic over time*, that is, the *direction* of authenticity is important. We are still a long way from reaching this level of understanding for most ecosystems. In areas where wholly natural forests have all but disappeared, as in much of Europe, ecologists are faced with a laborious process of building up knowledge from fragmentary ecosystems (Peterken, 1998). Even in places where greater areas of near-natural forest still exist, data collection and interpretation is a long-term challenge.

The issue of authenticity of ecosystems may be separate from questions of biodiversity richness or other desired qualities. Therefore, while it is clear that many social and economic aims will tend to encourage management systems that reduce authenticity, the same may also be true for some conservation aims. The extent to which conservation of supposedly natural areas will or should include conscious manipulation has as yet scarcely been addressed within the protected area and conservation communities; although many protected area managers have answered it by default and are consciously manipulating the ecology of lands under their control. Questions about whether or not to maintain high levels of authenticity in landscapes, or in a proportion of the landscape, will increasingly be addressed as conscious management decisions, at least partly determined through negotiation amongst the key stakeholders involved. Attempts to preserve the mix and proportions of species present at one particular moment in time are likely to run into difficulties or require continual and sometimes expensive management interventions. Increasingly, ecologists and land managers are recognizing the importance of maintaining the *process* within a system, that is, not just preserving species but also ensuring continuation of the ways in which they interact. The authenticity concept incorporates elements of both composition and ecosystem function in a framework for the assessment of the conservation value of the forest.

The main components of authenticity

Six major components have been identified as important in defining the authenticity of a forest ecosystem (drawing on Ratcliffe, 1993 but with modifications):

1 the *composition* of tree species and other forest-living plant and animal species – that is, an assessment of biodiversity;
2 the *pattern* of intra-specific variation, as shown in trees by canopy and stand structure, age-class, under-storey and so on;
3 the *functioning* of plant and animal species in the forest;
4 the *process* by which the forest changes and regenerates itself over time, as demonstrated by disturbance patterns, forest succession and so on;
5 the *resilience* of the forest in terms of tree health, ecosystem health and ability to withstand environmental stress;
6 the *continuity* of the forest with respect to area, edges, connectivity, fragmentation and age.

All these components are in turn affected by a seventh, related component:

7 *Development patterns*: authenticity can be deliberately suppressed or encouraged by management mimicking natural ecological processes, integration of forest into the landscape and so on, and is also affected by a wide range of other development patterns.

These components will be explained in greater detail below, along with some discussion about how they might best be measured. The three chapters looking at criteria all follow a similar format, with each component starting with a brief summary table and then discussing theoretical background and methods of measurement.

Authenticity component 1: Composition

Authenticity of composition is often used to determine conservation importance. Important indicators include the proportion of native and exotic species, the relative proportions of different native species and the absence of expected species.

TABLE 5.2 **Authenticity component 1: Composition**

Methods of data collection	Existing surveys and species lists, field studies, choice and recording of indicator species
Expertise needed	Basic understanding of biology and if necessary also field survey techniques
Likely costs	Literature survey low, field surveys can be high
Pros and cons	If no existing research exists this can be an expensive part of the survey but is a cornerstone of understanding about authenticity

BACKGROUND

All plant and animal groups are significant in the composition of an authentic forest and casual observations can be deceptive; forest with an authentic tree composition may for example have lost parts of its lichen and bryophyte population as a result of air pollution (Hawksworth and Rose, 1976; Tickle et al, 1995). Under some management systems particular microhabitats such as dead timber can virtually disappear (Dudley and Vallauri, 2004; Vallauri et al, 2005). As discussed above, many apparently natural systems have been profoundly altered and this affects composition. One of the reasons why management systems failed to address many biodiversity issues in forests, even in those countries where forests have been carefully managed for many years, is that they tended to focus on a few large species rather than looking at the overall quality of biodiversity.

It is frequently assumed that authenticity necessarily implies an extremely varied forest; this is often but not always true. Some forests are made up of a wide variety of species, and/or contain uneven-aged stands. Others may under natural conditions consist of virtual even-aged, monoculture stands, for example due to initial regeneration following major disturbances such as fire or windblow. Many forests have no fixed 'climax vegetation', at least on a site scale, but undergo a more-or-less constant cycling of species. Some forests are changing due to climate change. Authenticity is not necessarily equated with diversity.

Note: In the mid-20th century, forest conservation management in Sweden and Finland focused mainly on maintaining large mammals such as reindeer and elk. While this was successful, the focus on large mammals meant that other species were under-valued, so that for instance today Sweden has over 800 species associated with dead wood on its Red List of endangered species.

Source: Nigel Dudley

FIGURE 5.1 **Focusing on large mammals in Sweden and Finland**

MEASURING COMPOSITION – PROS AND CONS

This focuses on biodiversity surveys, use of existing species lists and in particular selection of key indicator species, which will sometimes include indicators of invasive species. If full surveys are possible, or have already been carried out, this information provides invaluable baseline data for the assessment.

If, as will often be the case, data are incomplete or lacking altogether, new surveys will be needed. Rather than attempt to identify everything within the ecosystem (a virtually impossible task) surveys will normally focus on a suite of species that together build up a picture of the overall composition. In Table 5.3 some generalized indicator types are suggested along with what they might tell.

Biodiversity monitoring is costly in terms of money, resources and skilled personnel. There is almost an inverse relationship between the richness of biodiversity and the amount that has been recorded. For instance, in many British counties, presence or absence of flowering plants is now mapped at the level of 1 kilometre squares, while in the Congo Basin there is still an area the size of Germany and France combined in which we do not even have information about presence or absence of elephants (information from WWF office in Gabon). New species, even of mammals, are still frequently being discovered, for instance in Borneo, Madagascar and Indochina.

Relatively little work has been done to draw together generalized assessments of the likely indicator values of groups and species. In practice, choice of indicators often reflects the particular interests of the specialists involved, rather than an attempt to provide an overall assessment.

Large mammals and birds are the most commonly measured indicator species, particularly in poorly studied forest areas. Selecting those that provide more general information – such as those dependent on old-growth characteristics like deadwood (Raphael and White, 1984), for example, woodpeckers in Scandinavia (Angelstam and Mikusinki, 1994) and North America (Bull and Johnson, 1995) or those at the top of the food chain (such as large birds of prey) (Tjernberg, 1986) – can help to

TABLE 5.3 **Some indicators of authenticity of composition**

Indicator	Type of measurement	Units and methodology
Authenticity of composition	Overall structure	Tree species and subspecies
	Presence and significance of composition of key groups	For example, flowering plants can indicate age of forest
	Indicators of particular microhabitats	For example, beetle species confined to standing dead trees
	Indicators of scale	For example, landscape species
	Presence of alien species	Number and status of invasive species
	Ecosystem	Presence of recognized ecosystem type(s)

build up a broader picture of likely biodiversity composition. If these species are present it suggests, although of course it does not prove, that others associated with the same type of habitat are likely to be present. Birds have been used as indicators in the tropics (Johns, 1996), including in comparisons between plantations and natural forests (Carlson, 1986). Birdlife International argues that diversity of birds is a surrogate for overall diversity (Stattersfield et al, 1998), which may be true but few rigorous comparisons have been made.

Although birds and mammals are useful because information about them is likely to be more readily available than for other species, they are not necessarily the best indicators, often being more adaptable to changing conditions than many other organisms. Large mammals, such as elephants and large apes, are relative easy to survey but are also often quite tolerant of degraded forests. More specific indicators, such as hornbills in much of the tropics, provide a more sophisticated picture but are correspondingly tricky and expensive to survey. Birds are also relatively ineffective indicators in disturbed habitat, where their response may be much less marked than other species (Dudley, 1992; Sallabanks et al, 2001). For example, pollution in some European forests is thought to be the factor leading to a rapid decline in spider diversity, but this has apparently had no impact on the diversity of the birds feeding on spiders (Clausen, 1986). Changes in bird communities as a result of management changes have been recorded in some situations for instance in Florida (Repenning and Labisky, 1985) and the UK (Ford et al, 1979), as have the impacts of forest fragmentation (Rolstad and Wegge, 1987) and other forms of management (Virkkala, 1987). Where data are scarce, careful interpretation of changes in birds and mammals can help to build up a generalized picture but must always be treated with caution.

In countries or regions where a relatively good understanding of biodiversity already exists, lower plant forms such as fungi (Bader et al, 1995), lichens (Söderström, 1988) and bryophytes (Gustafsson and Hallingbäck, 1988) have assumed an increasing importance in forest assessment, as their dependence on certain forest conditions has been noted and identified. In Northern Sweden for example, detailed indicators of forest richness have been devised using fungi and lichens (Karström et al, 1993) and in the UK, flowering plants have been used to describe changes under different management regimes (Kirby, 1990) and to identify relative ages of forests. Such detailed data are still a distant dream in many areas.

Care should be taken to avoid using more complex survey methods than are necessary. For example, using rare birds as an indicator of old-growth forest is probably a lot more complicated than carrying out a survey of deadwood – the latter does not fly away or hide – *unless* there is an existing survey

Note: The Wildlife Conservation Society has been developing the concept of *landscape indicator species*: that is, animals that through their range and space requirements help to define the minimum size of a particular landscape in terms of conservation values. At the other extreme, conservationists in northern Sweden use the presence of particular lichen species as a quick way of defining the oldest and most valuable forests from a conservation perspective.

Source: Nigel Dudley

FIGURE 5.2 **An African elephant at the edge of Ruaha National Park, Tanzania**

of birds, which can be analyzed to see what it says about forest condition. An ideal indicator species is one that will be fairly common *only* when a certain set of ecological conditions are present (Dudley and Jeanrenaud, 1997). Although use of experts can help reduce the time needed in surveys of composition, the specialists need not come from academia. Indigenous people and others living in close proximity and involvement with a particular forest system may have a much greater knowledge of key elements in biodiversity than an incoming scientist. This knowledge may focus on particular elements of use to human communities, such as medicinal herbs, food plants, conditions necessary for game animals and so on.

Many indigenous or local people have a high level of understanding about the ecological conditions needed by literally hundreds of different plant species. There is increasing experience in the scientific community about how such knowledge can be shared and used (Laird, 2002) and methodologies for facilitating this exist. The resulting knowledge gained can sometimes be particularly useful in the identification of richness of secondary forest fragments, as is the case of the *tembawangs* or fruit gardens developed by the Dayak people in parts of Borneo, which are now the nearest equivalent to natural forest in areas that have been heavily cut over, such as parts of West Kalimantan (Michon and de Foresta, 1995). Areas with a long history of habitation by human groups dependent on natural resources offer the best opportunities to use indigenous knowledge within surveys. It is important that incorporation of indigenous and local communities into surveys is done sensitively and with respect to their own rights and needs – codes of practice and guidance for this are emerging (for example Anderson, 2005)

Quite apart from their biological value, indicators will only be effective if users feel confident to collect and interpret the information that they contain and are convinced of their value. Indicators that can only be used by a few specialists have limited value, as do those that may cause misunderstanding or even antipathy in certain users. For example, people trained in forestry may be happiest with structural characteristics, while biologists and amateur naturalists may get positive pleasure from the time and effort needed to track down obscure species. Indicators also need to be carefully explained and put into context. The use of the spotted owl in the western US helped further polarize an already bitter debate by (wrongly in our view) suggesting that people are offered a choice between the life of one small bird and a logging community. The dangers of over-emphasizing indicators, rather than the whole ecological and cultural values of the forest, should also be noted.

SOURCES OF FURTHER INFORMATION

Published information or manuscript data on species probably needs to be tracked down at the site, although global repositories are starting to emerge such as the Global Biodiversity Information Facility (http://www.gbif.org). Surveying species present in a particular location is usually the first stage in any biological or ecological assessment, and many techniques already exist for capturing this information (for example Hawksworth et al, 1997; Sayre et al, 2000; Kapos et al, 2001). There is also growing expertise about the use of local surveyors and indigenous knowledge (for example Danielsen et al, 2000).

Authenticity component 2: Pattern

TABLE 5.4 **Authenticity component 2: Pattern**

Methods of data collection	Usually field surveys or aerial data information – some good satellite images also give an indication of forest pattern
Expertise needed	Basic knowledge of field survey or interpreting aerial data
Likely costs	Travelling and recording through the area, depending on size
Pros and cons	Important but hard to achieve without field work, unless existing survey data exist (which is rare)

BACKGROUND

A single-age plantation of a native mix of tree species might have an authentic composition but is not authentic in an ecological sense. The next stage in building a more complete picture of authenticity is by reference to the intra-specific variation and structural pattern of the forest (Ripple et al, 1991). *This includes both the pattern within a forest stand and the pattern that different forest stands make up within the overall landscape mosaic.* Patterns vary widely between forest ecosystems and will often need to be defined on a case-by-case basis, but might include for example:

- *Within the stand*: the concept of climax vegetation on a site scale – that is, a fixed vegetation type reached after a series of intermediary stages – has been largely disproved, however the idea of a *climax mosaic* of vegetation – that is, likely patterns of vegetation in a landscape, showing various influences and disturbances – remains of interest. Understanding the probable mosaic (for example the proportion of forest likely to occur in an old-growth status under natural conditions, the natural incidence of fire, regeneration patterns and disturbance patterns) would provide a powerful tool in planning managed forest landscapes. However, in many forested areas such knowledge is lacking and, where conditions have altered dramatically, it may be difficult to recover. Within stands pattern is reflected most clearly in the shape of the tree canopy and the presence and shape of under-storey vegetation, for example different sizes and ages of trees (Mladenhoff et al, 1993), tree canopies (Whitmore, 1990) and presence of a natural under-storey (Kuusipalo, 1984). Management or unplanned human interference tends to simplify forest pattern by for example creating uniform age classes of trees through felling or fire, or by removing under-storey vegetation through increased grazing pressure or over-collection. However, complexity does not invariably mean authenticity. Some forest stands will be naturally highly simplified, for example following fire, as is the case for Alpine ash forest in southeast Australia, while felling can conversely lead to a rapid increase in numbers of herbaceous plant species in western Canada (Hamilton and Yearsley, 1988) and the presence of dense brush in the Congo Basin. Site-level pattern indicators therefore need to be chosen on the basis of known vegetation patterns for a particular ecosystem.

• *Between the stands*: the mosaic of different vegetation types and ages at a landscape scale, including open space (August, 1983) and non-forest vegetation can all be important.

MEASURING PATTERN – PROS AND CONS

Many methodologies exist for assessing forest pattern or structure, at both site and landscape level, depending on time and resources available. Most of these will be too detailed for the purposes of a rapid, landscape-scale assessment. Quicker approaches will vary with the scale of the landscape and can range from use of aerial photography and satellite imagery to ground surveys of canopy pattern and microhabitats. A quick measure of site pattern is usually possible through a visual examination of the area. In most cases surveys will primarily be intended to identify those areas of forest where a reasonably natural pattern still exists rather than a particularly sophisticated assessment of small differences in pattern. Some attempts to ground-truth remote data should ideally also take place.

SOURCES OF FURTHER INFORMATION

Assessment systems have been developed for natural forests including tropical (der Steege, 1993) and temperate (Ripple et al, 1991), boreal (Pastor and Broschart, 1990) and mangrove (Snedaker and Snedaker, 1984) forest types, for managed forests (Ferris-Kaan et al, 1996) and as ways of comparing between the two (Swanson et al, 1990; Mladenhoff et al, 1993). Some examples of indicators are given in Table 5.5.

TABLE 5.5 **Some indicators of authenticity of pattern**

Indicator	Type of measurement	Units and methodology
Authenticity of pattern	Forest mosaic	Maturity classes of trees Volumes of timber Diameter of trunks (Spurr, 1952)
	Overall landscape pattern	Interpretation of GIS data (McCormick and Folving, 1998)
		Interpretation of skyline data to identify old-growth forests

Relatively little information is available about rapid surveys of forest pattern at a landscape scale because until recently such surveys have not been attempted.

Note: A survey of canopy shape can often identify patterns in forest ecosystems far more quickly than detailed surveys of particular indicator species. A rapid survey of canopy pattern, by someone who knows what is expected in a particular habitat, can quickly distinguish old-growth forests, logged over forests, plantations and secondary forests.

Source: Nigel Dudley and Sue Stolton

FIGURE 5.3 **Forest canopy patterns**

Authenticity component 3: Functioning

TABLE 5.6 **Authenticity component 3: Functioning**

Methods of data collection	Usually field surveys of particular indicators
Expertise needed	Basic knowledge of field survey techniques, in some cases detailed taxonomic knowledge depending on indicators chosen
Likely costs	Variable
Pros and cons	Great care needs to be taken in the choice of indicators to ensure that a broad picture of ecological functioning develops

BACKGROUND

The physical and chemical interactions that allow plant and animal species to survive and function in the forest extend the concept of authenticity beyond the species and the basic structure of a forest to the ways in which the forest ecosystem works over time, including:

- nutrient cycling;
- food chains and webs;
- relationships *between* species such as parasitism, symbiosis and commensalism;
- relationships *within* species such as territorial behaviour, social behaviour and intra-specific competition;
- soil relationships including chemical relationships such as allelopathy;
- presence of key microhabitats that indicate a healthy or complete ecosystem, including for example the presence of deadwood at varying stages of decomposition (Maser et al, 1988; Kirby, 1992a), or health and distribution of epiphytic mosses and lichens.

MEASURING FUNCTIONING – PROS AND CONS

Many of these processes are difficult to measure or identify in the field without a long-term research project, so surrogate indicators are particularly important. In the context it is being applied, measurement of ecological functioning is intended to give a broad idea about the extent to which natural ecosystem processes are taking place rather than attempting to understand or report on all these processes in detail.

Therefore, while functioning is in theory complex and potentially extremely time-consuming to measure, here it can be addressed much more crudely if effective surrogate indicators can be identified, which reflect the ecosystem. Some possibilities are listed in Table 5.7.

TABLE 5.7 **Some indicators of function**

Indicator	Type of measurement	Units and methodology
Authenticity of functioning	Viability of populations	Presence of indicator species
	Integrity of food webs	Presence of indicator species
	Site characteristics	Hydrological integrity Nutrient availability
	Continuity of forest	Maximum age of trees Period of continuous forest cover
	Presence of key microhabitats	Presence of dead standing and down wood Distribution of epiphytic species of lichens and mosses

The key aim here is to determine if the forest landscape is working well, both in terms of having a functional ecosystem (food web, chemical interactions and so on) and – much more difficult to determine – whether all the expected components of this ecosystem are in place. Many of the indicators chosen as indicators of composition can also give useful information here as well.

In Figure 5.4 we suggest presence of dead timber as one possible surrogate (although it should be noted that this is also a possible way of showing *process* in a forest ecosystem: many indicators can provide different types of information). Deadwood is a crucial and often under-represented microhabitat in managed forests. For example, in natural European broadleaved forest deadwood will eventually rise to anything from 5–30 per cent of the total timber, with volumes normally from 40 to 200 cubic metres per hectare and average volumes of 136 cubic metres per hectare (Christiansen and Hahn, 2003). Deadwood can rise even higher after a catastrophic event like a storm. These figures contrast dramatically with deadwood volumes in managed forests, even those that are managed in quite a natural manner. For instance, deadwood in the Jura Mountains of Switzerland, which are managed under continuous cover forestry with large areas in The World Conservation Union (IUCN) category V landscape protected area, was only 6.3 cubic metres per hectare (WSL, 2003). Less natural forests, such as plantations of Eucalyptus or spruce, result in a further significant reduction of volumes of deadwood (Elosegi et al, 1999). Species associated with deadwood now make up the largest single group of threatened species in Europe. For example, of the 1700 species of invertebrates in the UK dependent for at least part of their life cycle on deadwood, nearly 330 are Red Data Book-listed because they are rare, vulnerable or endangered (Smith, 2004). In Sweden,

Deadwood is a valuable indicator of function. In Finland, presence and type of deadwood are included in the Finnish Forest and Parks Service's forest inventories as an indicator for choosing which biotopes to protect. Normal forests contain approximately 5m3/ha deadwood although this sometimes increases to 10–40m³/ha. Dead and decaying timber is measured through use of 10 metre wide transects and dead trees classified according to criteria:

> species; standing or lying; amount; number of stems; length of dead tree; diameter of stem; state of decay (according to a three point scale); cause of death if known.

Several criteria help select valuable old-growth forest fragments:

> a certain percentage of deadwood; the structure of living trees; the absence of cutting since at least the 1930s; geological features; surveys of key indicator species.

Source: Nigel Dudley

FIGURE 5.4 **Standing and lying deadwood in Finland**

one of the most densely forested countries in Europe, 805 species dependent on deadwood are on the national Red List (Sandström, 2003).

Other relatively simple indicators for functioning might be presence of mycorrhizal fungi associated with certain trees, population of top predators (indicating a healthy food chain) and populations of soil micro-organisms.

Choice of indicators will depend on individual forest ecosystems, but we suggest including *at least one indicator to show a functioning food web and other indicators to show those components of the ecosystem liable to be absent or damaged.* Local expertise is important here. The extent to which this can be reliably measured will also depend on the area being assessed; over large areas information about food webs will probably have to be coarse and to some extent speculative.

SOURCES OF FURTHER INFORMATION

There is a lot of excellent material on ecological functioning and its measurement (for example Boorman and Likens, 1979; Oliver and Larson, 1990), which can help to identify useful indicators and techniques for assessments.

Authenticity component 4: Process

TABLE 5.8 **Authenticity component 4: Process**

Methods of data collection	Field surveys, we propose use of a simple classification typology; much of the information can often be gained from interviews
Expertise needed	Basic ecological expertise
Likely costs	Field surveys depending on area
Pros and cons	This is a critical element, but only in general terms and its measurement to the level needed here should not present a huge challenge

BACKGROUND

The next defining factor is the process by which the forest changes and regenerates itself, thus maintaining the ecosystem (Picket and White, 1985; Attiwill, 1994). 'Process' covers all aspects of regeneration including both gradual changes and response to catastrophic events such as storms and fires, including:

- gradual changes over time (Thompson, 1980; Runkle, 1982; Delacourt and Delacourt, 1997);
- regeneration patterns following catastrophic change (Pyle, 1997);
- regeneration patterns without catastrophe including recruitment patterns (Lorimer, 1989);
- changes as a result of environmental change, such as climate change (Hulme and Viver, 1998);
- tree longevity;
- direction of change within the forest.

In natural conditions, disturbance patterns play a critical role in defining the structure and functioning of the ecosystem (Attiwill, 1994). A large proportion of the world's forest regeneration patterns have been dramatically and persistently altered as a result of human activities. In many apparently natural forests (and even in areas set aside as nature reserves), management continues to alter the natural regeneration process. There is a significant proportion of professional foresters and protected area managers who believe that a forest will lose quality unless it is managed (a striking example of different perspectives on forest quality), and in large parts of the temperate and boreal forests the idea of leaving an 'over-aged' forest is anathema amongst the forest industry. Management tends to remove the oldest trees, simplify the age structure, and alter natural regeneration mechanisms, sometimes suppressing them while in other cases they are increased to unnatural levels. Forest fires fit into both these categories in different areas, for instance over-suppression of forest fires can have major impacts on forest structure and disadvantage those species that rely on fire for germination or to help them out-compete more fire-susceptible species (Baker, 1992). Climate change is today adding an important new factor by for example increasing the number of catastrophic weather events that change forest ecosystems (Markham et al, 1993).

Attempts to reverse this process have frequently been met with entrenched opposition; a recent example is the debate in North America about fires in forests and the 'necessity' of felling old-growth forests to remove fuel. In contrast, a growing number of conservation professionals and others are calling for various forms of 're-wilding' or 'wilderness recreation', where areas of forests and other natural vegetation are allowed to regain a fully natural ecological pattern (Soule and Noss, 1998; Taylor, 2005). In managed forests, the debate continues about the use of natural regeneration systems, although these are increasingly being investigated (de Graaf, 1986; Gomez-Pompa et al, 1991; Mansourian et al, 2005). In cases where forests are used for multiple purposes, are near human habitation, or are too small to accommodate the full mosaic of natural regeneration patterns, then some compromise between natural and human-controlled regeneration may be needed, even in areas ostensibly set aside to protect natural biodiversity and ecosystems. However, changes in human population patterns and management approaches also mean that some forests that have been managed for centuries or millennia are now being allowed to regain natural regeneration patterns; forests in protected areas in Australia, which would have previously been managed by aboriginal peoples through the use of fire, are an example of this (Flannery, 1994).

Our aim here is to help develop an understanding of the extent to which natural regeneration patterns may continue or have been altered through deliberate management or other accidental pressures. The management implications will almost always be the subject of debate.

MEASURING PROCESS – PROS AND CONS

The extent to which management has changed regeneration patterns creates particular problems in terms of measuring process; in many forests we have only a theoretical idea of what form a fully natural process of regeneration might take, while in others research over many years has helped understand these processes (for example Watt, 1925; Mount, 1973; Ward and Parker, 1989). Where forests have changed dramatically, comparison with other similar habitats can help. Scientists in Scandinavia had to rethink many ideas about forest regeneration when the collapse of the Soviet Union gave them access to the more natural forests on the other side of the border (H. Karjalainen (WWF Finland), 1994, personal communication). Table 5.9 outlines some of the information required.

TABLE 5.9 **Some indicators of process**

Indicator	Type of measurement	Units and methodology
Authenticity of process	Longevity of the forest	Age of oldest trees Presence of species indicating long-established or original forest
	Disturbance regimes	Presence of natural and unnatural disturbance regimes Presence and types of deadwood

Age is often a quick, if somewhat approximate, surrogate for process. In landscape mosaics where few if any trees of the maximum possible age remain, identifying the reasons for change can further strengthen the analysis. While natural regeneration patterns are complex and often poorly studied, for practical purposes they usually only need to be understood in fairly approximate terms and in many quality assessments a general picture of process is as much as will be necessary (for example to identify those areas where the most natural processes remain as likely conservation areas). A brief typology may be usefully applied to capture this information in a fairly qualitative form, as outlined in Table 5.10.

TABLE 5.10 **Outline typology for classifying process in forest regeneration**

State	Description
Fully authentic process	Forest landscape with trees of the maximum possible age and/or with fully natural regeneration patterns (signs of fire, wind blow and other changes without deliberate suppression)
Fairly authentic process	Forest landscape with some trees of the maximum possible age and/or with natural regeneration patterns but evidence of interference in regeneration (for example selective felling, changes in fire ecology, coppicing)
Process recovering authenticity	Forest landscape with evidence of past disturbance (for example no trees of the maximum possible age, lack of deadwood, evidence for past felling) but where management interference has now been withdrawn (seen for instance in many new protected areas)
Process with low levels of authenticity	Forest landscape with clear and widespread signs of management interference in the regeneration cycle (for example regular felling, fire suppression)
Non-authentic process	Forest landscape consisting of even-aged stands of trees, regularly managed

In Figure 5.5, this typology is applied to some temperate and boreal forests around the world. The typology is clearly approximate: in cases where more detailed information is required, further research will be necessary. It is also often difficult to understand process by simply looking at a forest (except in the case of extremely unnatural process) and some knowledge of forest history will also usually be required.

SOURCES OF FURTHER INFORMATION

Most discussion about forest process is at a research or theoretical level and simple field measurement systems do not exist, which is why we are proposing a new typology.

Note: Application of the typology for classifying process (clockwise, from top left):

- Fully authentic process: old-growth forest in the Bialowieza National Park Poland, unmanaged since the 1300s (Bobiec, 2002)
- Fairly authentic process: forest in national park in Arizona, US with natural processes but continuing fire suppression
- Process recovering authenticity: forest set aside to develop natural characteristics in Triglav National Park, Slovenia
- Process with low levels of authenticity: mixed oak woodland (grazed by sheep) and larch plantation in Wales
- Non-authentic process: Sitka spruce plantation in Wales.

Source: Nigel Dudley and Stephanie Mansourian

FIGURE 5.5 **Application of the typology for classifying process**

Authenticity component 5: Resilience

TABLE 5.11 **Authenticity component 5: Resilience**

Methods of data collection	Tree health surveys, information about likely pressures
Expertise needed	Some knowledge of botany and forestry
Likely costs	Costs of field surveys
Pros and cons	Unless studies have already been completed (which is unlikely in much of the world) this cannot be addressed without detailed field visits

BACKGROUND

To some extent this is a measure of forest health, although the parameters used are wider than in conventional health assessments, referring instead to the health of the whole ecosystem. Some degree of 'ill health' is expected under natural conditions in forests as a result of pest attack, disease and the normal process of senescence and will not affect a forest's ability to continue functioning (although they may be of concern to those interested in managing the forest for particular resources). More serious problems arise when ill health comes through human actions, for example introduction of exotic pests and diseases and the impacts of air pollution and climate change. (The 'tolerable' level of ill health also depends to a large extent on what a particular forest or wood is being used for and who is being asked to make the judgement.) Resilience includes reflection of the degree to which a forest can resist these and other stresses. Elements include:

- tree health;
- ecosystem health;
- presence of likely stress factors (pollution, pests and so on);
- trends in stress factors;
- ability to tolerate environmental stress.

MEASURING RESILIENCE – PROS AND CONS

Detailed methodologies exist for assessing some forms of tree health, including for example pollution damage (Schütt et al, 1983) and pest attack, although less is known about addressing this within a landscape context. What would be a tree health problem (and a low quality forest stand) to a professional forester interested in timber production might be a beneficial periodic dieback to an ecologist interested in natural disturbance patterns, and it is difficult to tell temporary health decline due to natural fluctuation from longer-term decline caused by our actions (Kandler, 1992).

While information on status indicators such as tree health can be incorporated if it exists, data on level of known threats may be a more useful surrogate at a landscape scale. Trend factors are potentially important in all the indicators discussed in this book, but they are perhaps particularly critical in the case of resilience. Table 5.12 outlines some possible indicators, including both status and threats.

Assessing pollution impacts on trees is a lengthy and controversial process and therefore outside the scope of a rapid assessment (see Box 5.2), whereas finding out if ambient pollution levels exceed their critical loads (that is, the level at which they are likely to cause ecological changes) is far easier in many parts of the world and will provide a quicker and probably more reliable measure of likely impact on resilience.

Use of the critical load concept can help in places where these are known. A critical load is the *quantitative estimate of an exposure to one or more pollutants below which significant harmful effects on sensitive elements of the environment do not occur according to present knowledge* (UNECE, 1988), that is, a measure of the damage threshold for pollutants. Critical loads have been set for a range of different habitats and species, to date mainly in Europe.

Scientists acting under the auspices of the United Nations Economic Commission for Europe (UNECE) have collated critical load data for sulphur and acidity levels throughout Europe, and have produced maps showing where the tolerance of soils and waters is already exceeded, or is likely to be exceeded in the future (Henricksen et al, 1992). Research carried out for World Wide Fund for Nature (WWF) pinpointed important European nature conservation areas likely to be at high risk from air pollution. Under controls proposed by the 1985 sulphur protocol, some 71 per cent of the protected areas studied are in areas suffering excess acid pollution. Even if countries were to adopt far more radical environmental scenarios, between 20–25 per cent of Europe's protected areas would remain at risk from acidification (Tickle et al, 1995).

While pollution levels linked to critical loads provide an ideal form of indicator, such precision is unlikely to exist for many ecosystems or stresses. Cruder indicators, such as the presence of invasive pests or high levels of pollutants will provide basic information about likely levels of threats and hence likely impacts on forest resilience. Climate change provides a particular challenge. While

TABLE 5.12 **Some indicators of resilience**

Indicator	Type of measurement	Units and methodology
Authenticity of robustness and resilience	*Status* indicators: Tree health	Scale and rates of mortality Tree health data (percentage leaf loss, dieback, increment etc.) Amount of salvage logging Number and scale of pest outbreaks or pest damage
	Other ecosystem indicators	Lichen health and health of other epiphytic species
	Threats indicators: threats to tree health	Levels of pollutants Introduced pests and diseases Unnatural levels of pests and diseases Introduced invasive species Quantity of pesticides and fertilizers applied Predictions of climate change impacts

general predictions may be possible for some forest landscapes (for example mangroves and some mountain top communities) resilience is likely to depend at least partly on factors such as the size and health of the existing forest (Noss, 2001). Good forest management during a time of changing climate differs little from good forest management under more static conditions, but with increased emphasis on protecting climatic refugia and providing habitat connectivity along environmental gradients, so that both assessments and any resulting prescriptions are likely to be generic.

SOURCES OF FURTHER INFORMATION

Many methodologies for measuring tree health exist including generalized methods for detecting air pollution damage (for example Schüytt et al, 1983) and for recognizing particular pest and disease attacks. Data on pollution levels are also important, as is information on other likely disturbance factors including invasive species.

Authenticity component 6: Continuity

TABLE 5.13 **Authenticity component 6: Continuity**

Methods of data collection	Maps, aerial surveys, satellite images
Expertise needed	Low
Likely costs	Low unless satellite images need to be bought
Pros and cons	Simple to complete on a superficial level but still very difficult to know the exact ecological implications of simpler biological corridors

BACKGROUND

This indicator covers the continuity of the forest in terms of both time and space giving a measure of the *period of continual forest cover* and the *area and degree of fragmentation of the forest today*.

Box 5.2   **Measuring forest health in Europe**

The United Nations Economic Commission for Europe (UNECE) and the Commission of the European Communities (CEC) carry out an annual survey of European tree health, working to conditions laid down in the *Convention on Long Range Transport of Air Pollutants*.

Symptoms vary between tree species, habitats and regions, and often also with altitude, a tree's location within a stand and sometimes according to the cause of decline. However, most symptoms are not specific to a single cause. In some cases, decline leads to death, and the condition becomes one of forest *dieback*. This remains comparatively rare in Europe, but is found for example in some areas of the Czech Republic, Slovakia and Poland. Despite the uncertainties resulting from these differences, some common symptoms can be identified:

- colour variations (especially chlorosis or yellowing) in leaves and needles;
- premature needle loss ('tinselling') or leaf fall;
- deformation in leaf shape and size;
- changes in the canopy of the tree, including thinning and the development of abnormal shapes, such as a condition in conifers called 'storks' nest';
- deformation in roots;
- abnormal branching patterns, including downward tilting of secondary conifer branches, known as the 'tinsel effect';
- disruption of natural regeneration;
- bark necrosis;
- susceptibility to disease and pest attack;
- reduced vigour and growth rate.

From this array of symptoms, the UNECE/CEC study uses four indicators in its annual survey of forest health in Europe:

1 degree of defoliation;
2 percentage of needle/leaf loss;
3 degree of discolouration;
4 percentage of discolouration.

This long-term data set is now the largest collection of comparative information on forest health in the world. Most researchers will have to make do with far less complete or accurate survey material.

To be authentic in an ecological sense a forest needs to be large enough to maintain genetically viable populations of all species, or be close enough to other forest to allow free interchange of species. It also needs to have been in existence long enough to retain, or to regain, a full range of species expected from the area. Factors include:

- area;
- edge type (that is, whether there is an artificially abrupt change or a natural gradation) (Ranney et al, 1981; Ferris-Kahn, 1991; Angelstam,1992);
- connectivity (Bennett, 1990; Mader et al, 1990);
- degree of fragmentation (Harris, 1984; Burkey, 1989; Esseen, 1994);
- age.

Some indicators are suggested in Table 5.14; it will be noted that these range from use of existing material such as historical records and maps through remote sensing to detailed site-level assessments. While very detailed assessments are possible the bulk of the information relates to age and area or shape of the forest with the main purpose to identify areas where scale and connectivity remain at natural levels.

MEASURING CONTINUITY – PROS AND CONS

Information is needed on both the physical area of the forest and its age. Area can often be calculated, at least approximately, through use of maps, aerial photographs and GIS data, and these sources are also increasingly utilized to measure fragmentation (Saatchi et al, 2001).

TABLE 5.14 **Some indicators of continuity**

Indicator	Type of measurement	Units and methodology
Authenticity of continuity	Area	Maps GIS (geographical information systems) data Aerial photography
	Age of forest	Historical records Tree ring analysis Analysis of pollen records
	Integration	Links between forests and other habitats Forest pattern Mapping
	Ratio of other land types	Mapping
	Connectivity	Presence of protected corridors and ecological stepping stones Protected area networks

In much of the developed world, age can be inferred through historical records, photographs and even reference to old paintings; in the absence of such data it can be calculated with varying degrees of accuracy from pollen analysis, tree ring analysis and from the use of indicator species of old-growth or ancient woodland (for example Kirby, 1988). In parts of North America, detailed indicators have been drawn up to identify old-growth forest, particularly in the Pacific Northwest where debates about felling old-growth have gained enormous political momentum (Franklin et al, 1981; Martin, 1992). Such methodologies are becoming more widely available, particularly in temperate and boreal countries. Assessing the impacts of fragmentation is likely to be more time-consuming although in recent years a number of experimental fragmentation indices have been developed and field tested, including one developed by the United Nations Environment Programme World Conservation Monitoring Centre (Kapos et al, 1997). The World Resources Institute (WRI) also defined criteria for defining what it has named 'frontier forests', which are those forests of sufficient age, naturalness and geographical extent that they are considered to be ecologically viable (Bryant et al, 1997).

SOURCES OF FURTHER INFORMATION

Data on the rough size and shape of forests and on their connectivity is becoming easier to access all the time. For many parts of the world, information freely available (for instance on GoogleEarth) will be good enough to gain an approximate idea of the links between different forest patches (and it also throws in stark relief how isolated many protected areas really are).

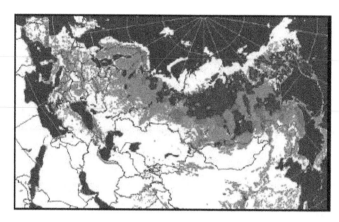

Note: The World Resources Institute (WRI) used a simple methodology drawing on forest area and degree of fragmentation to identify what they called 'frontier forests': those forests that remain in a near natural state of sufficient extent to be ecologically viable.

Source: Bryant et al, 1997

FIGURE 5.6 **Map of Europe and Russia**

Authenticity component 7: Agents of change

TABLE 5.15 **Authenticity component 7: Agents of change**

Methods of data collection	Information and analysis of pressures
Expertise needed	Various methodologies exist, and training in whichever one is chosen may be necessary
Likely costs	Data collection, possibly the costs of a workshop
Pros and cons	Easy to miss important factors (stakeholders may also have missed significant issues)

BACKGROUND

The last set of indicators is radically different in form, representing likely agents of change in authenticity, either increasing or decreasing the overall naturalness of the forest's ecosystem. There is some overlap here with resilience; the difference being that in the case of resilience, threats are likely to be distant (for example remote sources of air pollution) or accidental (for example introduced pests), whereas development patterns include changes consciously made within the landscape under consideration. In many cases, human disturbance patterns will tend to decrease authenticity – for example the development of roads, new settlements, agriculture, mining and various other forms of human impact. However, in other cases (for example during forest landscape restoration or as a result of changes in forest management practice), overall authenticity can be deliberately enhanced. Associated management factors are therefore critical to determining level and trends in authenticity. Some experience has developed with respect to maximizing authenticity within managed forests (for example Kohm and Franklin, 1997). An outline of indicators is given in Table 5.16.

TABLE 5.16 **Some indicators of agents of change**

Indicator	Type of measurement	Units and methodology
Development impacts	Presence of human disturbance around and impacting on the forest landscape	Roads, railways and other transport links Settlements Land-use patterns
	Deliberate interventions within the forest	Management in the forest (proportion of plantations, multiple-use forests etc.)

MEASURING AGENTS OF CHANGE – PROS AND CONS

In many cases measurement will simply consist of listing, identifying the likely impacts of such agents of change and mapping their impacts within the landscape. However, a series of more detailed methodologies exist, looking at both immediate causes and underlying causes of change.

SOURCES OF FURTHER INFORMATION

Much of the information on threat analysis in this context has been developed for protected areas but is transferable. Examples include the threat analysis developed as part of WWF's Rapid Assessment and Prioritization of Protected Area Management (RAPPAM) Methodology (Ervin, 2003). At a larger and more general scale, root causes analysis also developed by WWF (Wood et al, 2000) and the analysis in The Nature Conservancy's Five-S Framework for Site Conservation give approaches to capturing this information (The Nature Conservancy, 2000).

Measuring authenticity at a landscape scale

Within forest quality assessment, authenticity needs to be measured at a landscape scale. Most indicators – including many of those described above – have been developed for sites and will have to be amalgamated to represent landscape values. This is possible in the case of a major research project but will in many cases be too time-consuming or expensive. Authenticity at a landscape scale presents particular challenges and we include some proposals for how this might be tackled. Below we present three options; all rely on approximations rather than comprehensive surveys, although the latter still represent an ideal if time and resources allow:

1 a coarse level *typology* for categorizing forests by their level of authenticity as a first 'filter' in determining authenticity where detailed assessment systems are impracticable;
2 a simple *site-level assessment method* that can provide information on a variety of sites, which can then be drawn together to provide a landscape scale picture;
3 a simple *matrix* for presenting site-level assessments that can be amalgamated into (or extracted from) a wider landscape analysis.

A TYPOLOGY OF FOREST AUTHENTICITY

Summarized in Table 5.17, this describes a scale from 1–5 for degree of forest authenticity. Due to the complexity of forest ecology and the range of changes that can occur, these measurements will inevitably be approximate. The stages *very high* and *very low* are extremes with most forest types falling within the middle three categories. Such a typology could be used for a very rapid classification, perhaps using existing maps, to identify and locate those forests of the most importance from an ecological perspective (and also identifying forests that would require further study).

TABLE 5.17 **Typology of authenticity**

Stage	Descriptor	Details	Examples
1	Very low	Few natural species or ecological functions. Narrow range of seral stages and simplified structure	Monoculture or near monoculture exotic plantation, for example oil palm in Malaysia, *Pinus radiata* in New Zealand or eucalyptus in Portugal
2	Low	Highly modified forest. Limited range of possible species, often exotics present. Narrow range of seral stages. Limited size of trees and few old trees	Young bush resulting from regular cut and burn, intensively managed timber with planting or weeding, or heavily grazed relic woodland, for example swidden agriculture in West Kalimantan or intensively managed timber plots in Norway
3	Medium	Reasonably natural forest but with some components highly modified; variable size and continuity over time[1]	Forests managed extensively for timber production, 'forest gardens' or coppice, for example the Black Forest in Germany, forests in the Congo Basin affected by the bushmeat trade, forest gardens in Sri Lanka, Kalimantan and Sumatra and oak coppice in England
4	High	Forests approaching the natural state but with some key elements reduced or missing – for example oldest forest or some species[2]	Forests with a single selective logging (or logging long ago), natural forest fragments too small to support full biodiversity, for example logged over forest in Mount Kenya National Park, regrowth on abandoned farms in Costa Rica or Vermont US, recovering forest in Coramandel, New Zealand, continuous cover forestry in the Swiss Jura
5	Very high	Near natural forests with little human disturbance or management; all seral stages present or potentially present	Forests in protected areas or in areas that are currently either un-managed or managed without significant impact on natural structure and ecology, for example Tasmanian Wilderness Area, Australia, forests in the Western Ghats of India, boreal forests in Kamchatka, Russia and Central Amazon forests

1 Such forests could contain exotic species if they were well established and playing an ecological role; for example sweet chestnut (Castanea sativa) was probably introduced into the UK from southern Europe; however old English chestnut woods can contain a high biodiversity and a close to natural structure.
2 The distinction between 'medium' and 'high' authenticity is probably the most difficult. Can forests managed for commercial timber harvests ever have a high authenticity?

A SIMPLE SITE-LEVEL ASSESSMENT OF AUTHENTICITY

The second option aims to provide slightly more detail, assuming analysis at a stand level but without the need for detailed field surveys or recording, thus reducing time required. It could be used on stands throughout the landscape to gain an overall picture of landscape-scale authenticity.

Data card for stand-level assessment of forest authenticity:					
Indicator	**Elements**				
Composition	How natural is composition of tree species?	Fully natural	Mainly natural	Many exotics	Exotic
	How natural is composition of other species?	Fully natural	Mainly natural	Many exotics	Exotic
	Are alien species present?	Significant or invasive aliens	Some non-invasive aliens	No significant aliens	
	Overall authenticity of composition	Fully natural	Mainly natural	Significant exotics	Almost all exotic

Notes on composition

Pattern	What is the tree age distribution?	Mixed including old	Mixed middle age	Mixed mainly young	Trees recently lost	Mono-culture
	Is the forest canopy natural?	Fully natural		Mainly natural		Mainly unnatural
	Size of the forest in hectares					
	Overall authenticity of pattern	Fully natural	Mainly natural	Significant alteration	Monoculture

Notes on pattern

Functioning	Are viable populations of resident plant and animal species present?	All viable	Most viable	Many not viable	Most not viable
	What are the soil characteristics?	Stable		Limited erosion	Serious erosion
	What are the hydrological characteristics?	Healthy		Limited problems	Serious problems
	How much deadwood is present?	Natural amounts		Limited amounts	Virtually none
	Overall authenticity of functioning	Fully functioning	Mainly functioning	Significant loss of functioning	Not functioning in a natural way

Notes on functioning

Process	Does a natural disturbance regime exist?	Wholly natural	Partly natural	Mainly unnatural	Wholly unnatural
	Does an unnatural disturbance regime exist?	List unnatural disturbance factors			
	Overall authenticity of process	Wholly natural	Partly natural	Mainly unnatural	Wholly unnatural

FIGURE 5.7 **Stand-level scorecard for authenticity**

Notes on process					
Continuity	Age (approximate length of continuous forest cover)				
	Are the forest edges natural or artificial?	Wholly natural	Mainly natural	Quite unnatural	Wholly unnatural
	Is the forest connected to other similar habitat?	Wholly connected	Still well connected	Some limited connections	Isolated
	Overall authenticity of continuity	Wholly natural	Partly natural	Mainly unnatural	Wholly unnatural
Notes on continuity					
Resilience	What is the average tree health?	Good		Average	Poor
	What is the health of other environmentally sensitive species?	Good		Average	Poor
	Are there important introduced pests, diseases and invasive species?	List those that affect ecosystem health			
	What are air pollution levels?	High		Medium	Low
	Overall authenticity of resilience	Wholly natural	Partly natural	Mainly unnatural	Wholly unnatural
Notes on resilience					

FIGURE 5.7 **Stand-level scorecard for authenticity** *(continued)*

The data card, as shown in Figure 5.7, can be filled in quite quickly for individual forests or forest sites and many data cards can be assessed to build up a picture of forests in a landscape, although this presupposes either a limited scale of landscape or assessment of a relatively small proportion of forest sites as examples.

A MATRIX OF AUTHENTICITY

Typologies and data cards can be helpful in allowing a rough guide to authenticity to be drawn up on a landscape scale. However, they are inevitably approximate and also mean that all the components of authenticity have to be amalgamated within a single scale point. A more detailed and information-rich assessment could be developed by use of a matrix, where each of the components is scored separately and represented diagrammatically in Figures 5.8 to 5.10. As with the data card, such an approach is only suitable for a fairly small landscape or for a few examples within a larger landscape. In practice, it will probably be necessary to define each score in quantitative or at least descriptive terms, although these will vary from one ecosystem or situation to another.

Information could be represented in the form of either simple ticks within the matrix or as a bar graph as illustrated in the theoretical example below, for a small fragment of secondary forest managed for timber.

Component	Very low	Low	Medium	High	Very high
Natural composition					
Natural pattern					
Natural function					
Natural process					
Area/connectivity					
Resilience					

FIGURE 5.8 **Matrix of authenticity**

Component	Very low	Low	Medium	High	Very high
Natural composition			✔		
Natural pattern				✔	
Natural function			✔		
Natural process		✔			
Area/connectivity	✔				
Resilience		✔			

FIGURE 5.9 **Ticked matrix**

Component	Very low	Low	Medium	High	Very high
Natural composition					
Natural pattern					
Natural function					
Natural process					
Area/connectivity					
Resilience					

FIGURE 5.10 **Bar graph matrix**

Matrices of this kind might be prepared for forests within a landscape as part of an overall assessment. Examples of these approaches are given in Case Study 6.

AUTHENTICITY AND FOREST MANAGEMENT

In natural forests, semi-natural forests and multi-purpose forests, authenticity can play a role in helping set the framework for management policies. In particular, it can help to:

- identify priority sites for a protected area network (through integration with concepts of High Conservation Value Forests (HCVF) see Appendix 3);

- set management policies in managed natural or semi-natural forest;
- improve the ecological value of secondary or disturbed forests;
- design strategies for forest landscape restoration.

From a conservation perspective, a key assumption is that authenticity is more important at the level of landscape than for individual stands. However sophisticated management becomes, it cannot exactly duplicate the natural ecological processes and does not eliminate the need for large enough areas of forest to be set aside to sustain both species and ecological processes. Nonetheless, the *concept* of authenticity could serve as a backdrop against which to measure changes in management policy at a landscape level, the direction of future forestry developments and the principles of multi-purpose forestry.

With respect to management for conservation, authenticity should be a general aim rather than a straitjacket. In some areas, centuries-old forestry practices have developed an associated biodiversity of their own, even though they are 'unnatural' in ecological terms. Other conservation priorities may justify halting natural forest succession, for example to maintain grassland ecosystems such as African savannah (many of which are almost entirely cultural landscapes). Indeed, many of the world's apparently 'wild' areas have been managed for centuries or even millennia, often through the use of fire, cutting, selective grazing or the maintenance of wild game herds. Whether or not such ecosystems can now survive without human interference is in many cases a matter of conjecture, particularly when space for conservation areas remains limited.

The extent to which an area set aside for biodiversity conservation should be left to natural ecological processes should therefore be an early management decision. In many cases, questions about how much both conservation objectives and the aims of wilderness protection are actually linked to authenticity of ecosystems have hardly started to be addressed.

6 | Environmental Benefits of Forest Quality

In Costa Rica, we travel along a dirt road through torrential rain. We are sitting in the back of an open truck and our clothes are soaked by the time we get to our destination, deep down in a forested valley. The owner of a small hydroelectric station is there to greet us; his power plant generates 3 megawatts of electricity to supply the local town. He's made a success of this business and is proud of the station, which is in excellent repair. All around the land is settled by poor farmers and the power plant owner pays them a set sum every year not to clear the forest in order to protect his water supply. He knows that there is a debate about how much deforestation would alter the water flow, with some specialists thinking it would make little difference, but he doesn't care. Right now, there is forest and there is water and the amount that is paid to the farmers every year is a very small insurance policy. Further north in Guatemala, Pepsi Cola is also paying money to keep forest above one of its bottling plants, this time to maintain the exceptionally high water quality that will disappear if the forests are replaced by farmland and cattle.

The second criterion relates to *both human and non-human values*. Forests help to maintain ecological balance, which has major and very practical impacts on human societies. Although these roles are increasingly recognized and techniques have allowed them to be valued in a number of ways including by economic criteria, this understanding is still partial and in practice many values are unrecognized. There also remains a reluctance to pay for these services. A corporation that builds a factory in an area because of the abundance of clean water does not usually expect to pay to maintain the forest that, in turn, maintains the flow and purity of the water. Such values are generally seen as free. However, attitudes are gradually changing and many governments are now committed to the idea of maintaining environmental services from forests through their signature to international treaties such as the Convention on Desertification and the Framework Convention on Climate Change. Measurement of environmental benefits thus assumes a greater importance and in some cases a direct economic rationale. The following section summarizes some of the key issues and their related indicators at a landscape scale, including:

- biodiversity and genetic resource conservation;
- soil and watershed functions;
- impacts on other natural and semi-natural habitats;
- influence on climate.

Environmental benefits component 1: Biodiversity and genetic resource conservation

BACKGROUND

Forests contain more biodiversity – in terms of ecosystems, species and genetic variation – than any other terrestrial ecosystem. Many species have yet to be identified and described by scientists (Whitmore and Sayer, 1992).

A number of aspects relating to biodiversity conservation are covered by the first component of authenticity, which discusses composition of species in the forest. However, *quality* of biodiversity is not necessarily equated directly to either authenticity or to numbers of species, and the current

TABLE 6.1 **Environmental benefits component 1: Biodiversity and genetic resource conservation**

Methods of data collection	Use of existing databases, indigenous knowledge and possibly some field research
Expertise needed	Basic biological knowledge
Likely costs	Low for some surveys (for example presence and extent of protected areas), higher for research into economically important species etc.
Pros and cons	Measuring several things in one indicator can be complicated

component considers in particular the benefits that biodiversity provides to human society (Perry, 1993).

Forest biodiversity has utilitarian human benefits in that it provides us with a range of products – food, medicines, fuels, manufacturing materials, essential oils and others – including many waiting to be discovered or described (Freese, 1997). Forests sometimes directly support beneficial species and many cultural forest landscapes have been shaped at least in part with an aim of maintaining beneficial plants and animals (including some forest landscapes widely believed to be 'natural'). Species therefore may not play a critical role in the functioning of a particular forest ecosystem but have high biodiversity significance because they are rare, specialized in their distribution or endemic to a particular locality, or play a vital role in human well-being.

In addition to these utilitarian considerations forest biodiversity also has intrinsic ecological and existence values and many people believe in consequence that we have a moral obligation to prevent unnaturally high levels of biodiversity loss that may occur as a result of human activities. The role of forest management in maintaining forest biodiversity is recognized (FAO, 1993). Biodiversity conservation in forests usually requires two main elements of land management: setting aside a proportion of the forest estate primarily for biodiversity conservation, and taking account of biodiversity needs in the management of the remaining forest area, including where necessary restoration.

The amount of land in protected areas will determine the importance of management in the remaining forest; in areas where relatively little of the forest is protected then management in the remaining area will become increasingly important in terms of maintaining biodiversity. Today, emphasis is increasingly being put on the need for an *ecologically representative* system of protected areas (Dudley et al, 1997; Dudley and Pressey, 2001), and techniques such as gap analysis are being developed to draw up protected area strategies (Scott et al, 1993; Iacobelli et al, 1994). The minimum useful size for *total* ecosystem protection is often determined by the needs of large predators (Sanderson et al, 2002). Smaller protected areas can play critical roles in the preservation of particular species, or in a landscape in which larger animals have already disappeared. Including biodiversity needs in managed forests is an increasingly important element in many forest management plans. In areas where native woodlands have already been severely depleted or eliminated and there is a managed or cultural landscape, it may be necessary to preserve what are effectively 'unnatural' forest areas if specific rare species are to be protected (Bowden and Hoblyn, 1990). Indeed, in the highly managed forests of Europe, some adaptations to managed forests have now occurred within ecosystems, and these must also be taken into account when planning management strategies.

MEASURING BIODIVERSITY – PROS AND CONS

Many indicators will already have been covered in the section on authenticity composition in Chapter 5. Here, in line with the emphasis of this section on the interaction between human and non-human values, the emphasis of indicators is on three key aspects, as outlined in Table 6.2.

1   presence of rare or endangered species;
2   presence of species of direct value to humans (see also Social and Economic Component 2 in Chapter 7);
3   presence of efforts to protect biodiversity especially through protected areas.

A growing number of tools and methodologies exist for determining the importance of biodiversity to local human communities. Management implications will be affected to a large extent by the range of species required. In subsistence communities, literally hundreds of species of plants and animals may be utilized while in poor rural communities there may be a focus on a few with economic or cultural value, such as hunting of large game. Depending on the amount of information already available, such assessments may therefore be quantitative, if data on populations and values are known, or qualitative in cases where lack of information means that importance has to be ranked by a simple scale.

Broad assessments of biodiversity are becoming increasingly common, particularly as a result of various ecoregional planning processes by organizations such as World Wide Fund for Nature (WWF), Conservation International and The Nature Conservancy. These can provide a basis for making judgements within landscapes, although further amplification and refinement of information may be needed.

SOURCES OF FURTHER INFORMATION

Include Red List data (IUCN, 2000), increasingly sophisticated methods for scientific assessment of biodiversity (Boyle and Boontawee, 1995; Harper and Hawksworth, 1995; Miller and Lanou, 1995; and Morris et al, 1999), participatory methods for assessing the role of community forestry (Davis-Case, 1990; Messerschmidt, 1993; Carter, 1996; Jackson and Ingles, 1998;) and techniques aimed at working with local communities in assessing biodiversity values (Ogden, 1991; Campbell and Luckert, 2002). Assessment methods are beginning to aim at a landscape scale (for example Sedaghatkish, 1999). The Global Biodiversity Information Facility (http://www.gbif.org/) will again be a valuable source of information.

Basic data on protected areas can be accessed through the World Database on Protected Areas (WDPA, 2005) and in many countries also through nationally held databases, which will usually include geo-referenced data. As yet no global source of information on management effectiveness of protected areas exists but various methodologies for assessment exist (Hockings et al, 2000), including various rapid assessment methodologies (for example Ervin, 2003; Stolton et al, 2003).

Data relating to biodiversity under this section might include lists of priority species, including those that are unique and / or rare and those of value to human communities, along with their status, plus lists and maps of areas that have been set aside for biodiversity, either as fully protected areas or through other forms of management (hunting reserves, sustainable use areas, areas set aside for other uses that also have a biodiversity value such as protection for watersheds).

TABLE 6.2 **Some indicators of biodiversity**

Indicator	Type of measurement	Units and methodology
Rare biodiversity	Presence of important biodiversity	IUCN Red List of threatened species Other threatened species Endemic species
		Absence of key groups (for example those associated with fire, deadwood, old trees)
Utilitarian species	Presence of species of direct human value	Food species Medicinal species Fodder Species with saleable value etc.
Efforts to conserve species	Presence of protected areas	Details on area, shape, IUCN protected areas categories and incentives for protection Type of protection (government, NGO, private, community etc.)
	Effectiveness of protected areas	Degree of protection Presence of management plan Infrastructure
	Other forms of protection	Hunting bans Presence of set asides etc.

Environmental benefits component 2: Soil and watershed protection

TABLE 6.3 **Environmental benefits component 2: Soil and watershed protection**

Methods of data collection	Local stakeholders including water companies, possibly direct measurements
Expertise needed	Little specialized knowledge needed for basic information gathering
Likely costs	Low
Pros and cons	Still major debates about the role of forests in hydrology, which differ with type and age of forest, climate, soils etc.; care needed to avoid simplified or misleading answers

BACKGROUND

Natural forests, and well-managed secondary forests, provide important benefits in terms of conserving soil, regulating water flow at a local level in catchment areas and particularly for maintaining water quality. Conversely, poor forest management can lead to significant hydrological impacts.

The relationship between forests, soil and water has been recognized for hundreds if not thousands of years. Early efforts at reforestation, in Japan 500 years ago (Ministry of Environment, 2002) and in the Alps (Küchli, 1997) and Scandinavia (Ekelund and Dahlin, 1997) during the 19th century, were stimulated by problems of erosion and flooding caused by forest loss. Since serious soil erosion occurred in the US in the 1930s – a social and environmental disaster described in John Steinbeck's 1939 novel *The Grapes of Wrath* – the relationship between soils and forest management has played an important role in US policy (Binkley and Brown, 1993). Concern remains about the impact of fast-growing tree species on tropical soils (FAO, 1980).

Impacts on soil and water quality have direct ramifications for biodiversity and for wider forest values. In the US, the executive director of the Association of Professional Fish Biologists was quoted in the press as saying that increased logging in the US Pacific Northwest would contribute to 'the decline and extinction of native fishes over vast portions of their range' (Durbin, 1991). Studies in the early 1990s showed that within the Columbia River basin some 76 native salmon populations were at high or moderate risk of extinction due to logging and deforestation (Nehlsen et al, 1991). Conversely, the presence of coarse woody debris in forest streams in natural old-growth forests can increase spawning success (Spies et al, 1988) by creating gravel bars and pools which reduce water flow, create fish habitat (Bilby and Bisson, 1998) and provide valuable substrate for algae (Miller et al, 2004).

Deforestation sometimes causes an increase in total water yield in catchments as the deep-rooted trees are replaced with shallower rooted grasses or annual crops. This can cause a rise in the ground water tables and if salts occur in the soils they can be brought to the surface, killing the vegetation. In the south west of Western Australia, more than 400,000 hectares of arable land have been lost to production by rising saline water tables as a direct result of clearing eucalypt forests and woodlands during the past 100 years (McFarlane, 1991). Forest management operations, including construction of roads, can also create increased soil erosion unless carried out with care (Swanson and Dryness, 1975). Conversely, loss of tropical cloud forests (Bruijnzeel, 1990; 2001) and certain kinds of old-growth eucalyptus forest in Australia (Langford, 1976) can create a net reduction in water availability.

The presence of healthy forests is increasingly recognized as an important factor in maintaining drinking water supplies in many countries. In Puerto Rico for example, half the island's drinking water comes from the Caribbean National Forest area, even though this covers less then 3 per cent of the total area of the island (F. Wadsworth, 1997, personnal communication). A third of the world's largest cities draw a substantial proportion of their drinking water from protected forest catchments (Dudley and Stolton, 2003a), including Tegucigalpa in Honduras, Sydney in Australia, Los Angeles and New York in the US and Dar es Salaam in Tanzania. Forests help to maintain the purity of water sources, thus providing major reductions in the costs of water purification (Aldrich et al, 2000). The economic potential of water services from forests is increasingly being recognized and addressed through innovative financing mechanisms (Johnson et al, 2001). These benefits are known to be enormous. A recent study calculated that the presence of forest in Mount Kenya National Park saved Kenya's economy more than US$20 million through protecting the catchment for two of the country's main river systems, the Tana and the Ewaso Ngiro (Emerton, 2001). Projects using water resources as a springboard for various forms of payment for environmental services (PES) schemes have been most thoroughly developed in Latin America. In Costa Rica for example, the government has been involved in a scheme to help users such as hydropower companies to pay farmers to maintain forest cover in watersheds (Rojas and Aylward, 2002), while in Quito, Ecuador, water companies are helping to pay for the management of protected areas that are the source for much of the capital's drinking water (Pagiola et al, 2002). These initiatives have also increased our knowledge of how to measure the benefits of such systems.

However these relationships are complex. There is not necessarily a direct link between forest loss and soil loss – although this has often been claimed. Erosion from the Himalayas in Nepal for example is now thought to be due as much to natural erosion patterns as to deforestation, although in the past forest loss has been widely blamed for high sedimentation in rivers such as the Ganges (Hamilton, 1987). Although forest management has clear impacts on hydrological systems in North America, these are also not as simple as has sometimes been claimed (Grant, 1990). A recent survey also found few major impacts within the UK (Worrel and Hampson, 1997). It follows that measuring these impacts requires care and skill. Less controversially, riparian forests play an important role in reducing river erosion and regulating floods and the loss of many flood plain and river island forests has increased the impact of flooding in parts of Europe and Asia.

MEASURING SOIL AND WATERSHED VALUES – PROS AND CONS

While values such as water flow and water quality are relatively easy (if somewhat expensive) to measure, relating these directly to particular changes in management or particular events within a landscape can prove more difficult (exceptions may be when dramatic changes in forest cover provoke major alterations in hydrology, although interpretation of even these events remains controversial). A shorthand way to address this issue is to focus on evidence of active watershed protection and observable indicators of water quality; careful interpretation will in many cases be needed to link these indicators directly to overall forest status and quality. Some possible indicators are outlined in Table 6.4.

TABLE 6.4 **Some indicators of soil and watershed values**

Indicator	Type of measurement	Units and methodology
Watershed protection	Official protection	Areas set aside specifically for watershed protection
	Traditional protection	Areas managed by traditional peoples as communal fishing grounds
Soil and water quality	Evidence of quality	Soil loss (visible erosion, turbidity in water courses) Water quality Fish stocks Presence of other aquatic indicator species
Economic benefits	Value of watersheds in terms of provision of high quality drinking water	Replacement costs by other forms of purification

SOURCES OF FURTHER INFORMATION

Many technical methods exist for measuring water flow (Davis and Hirji, 2003a; Dyson et al, 2003) and water quality (Davis and Hirji, 2003b). Methodologies for calculating benefits also exist (Pagiola et al, 2002) although there is still considerable disagreement about precisely what the benefits are likely to be (FAO and CIFR, 2005), making assessment more difficult.

Environmental benefits component 3: Impacts on other natural or semi-natural habitats

TABLE 6.5 **Environmental benefits component 3: Impacts on other natural or semi-natural habitats**

Methods of data collection	Usually general, qualitative analysis of major impacts
Expertise needed	Basic understanding of environmental interactions, pressures and connectivity
Likely costs	Low to high
Pros and cons	This is almost like an Environmental Impact Assessment and can be as simple or detailed as necessary. Many sources of information exist

BACKGROUND

Forests do not occur in isolation but exist in a wider landscape containing a range of other habitats, all of which interact. As a result, many aspects of forest management, including choice of protected areas and of management regime, have important impacts on other habitats, including freshwaters, coastal areas (from mangroves), shrub and heath and peat (Safford and Maltby, 1998), with knock on effects for both biodiversity and humans. In addition to measurable hydrological impacts, summarized in the previous component, forests act as important buffers for coastal zones and forests and other wooded land, and interact with a range of savannah and tundra habitat types.

The most immediate effect is often on freshwater systems running near or through forests. In addition to the impacts on aquatic life described earlier, forests in catchments can have impacts on mammals and birds associated with freshwaters. Planting of conifer trees right up to the edge of streams had an adverse impact on otter populations in Wales for example and, as a result, wider buffer zones were developed. Forests also have an important ecological role in many coastal areas. Loss of mangroves in tropical regions can have a catastrophic impact on coastal fish populations because the mangrove roots provide nutrient-rich breeding grounds for many species – in Malaysia for example at least 65 per cent of harvested fish are associated with mangrove habitats (Linden, 1990). Ninety per cent of commercially important Indian fish and shellfish species spawn and breed in mangroves. In Queensland, Australia, mangroves were estimated to be worth GB£1500 per hectare per year for fish production (Crisp et al, 1990).

Poorly planned afforestation can also damage important semi-natural habitats. In the UK, semi-natural habitat such as moor has become an important habitat for birds such as the hen harrier (*Circus cyaneus*) and golden plover (*Pluvialis apricaria*), and planting with exotic conifers results in net loss of biodiversity (Ford et al, 1979). Similar debates are underway about plantations in Southeast Asia and Latin America, although in this case it is often natural forests that are being replaced by plantations (Carrere and Lohmann, 1996). At a landscape scale, methods for assessing the likely impact of afforestation schemes are as important as ways of looking at the consequences of forest loss, but have generally received less attention from researchers and academics.

MEASURING IMPACTS ON OTHER HABITATS – PROS AND CONS

In many cases this component will have to be reflected in general, qualitative terms, although precise data may be available (for example about likely impacts on fish stocks), either for a particular site or for similar habitats.

TABLE 6.6 **Some indicators of interactions with other habitats**

Indicator	Type of measurement	Units and methodology
Interactions with other habitats	List of the main interactions	List (and if possible statistics) of functions, species and goods and services linked to forests
Impacts from forest change (gain or loss)	Changes in ecological functions and/or associated goods and services	List (and if possible statistics) of functions lost and gained

To some extent, the purpose of this indicator is to ensure that wider ecosystems impacts are recognized and can be incorporated into any planning or implementation phase. A simple matrix might be the most effective way of summarizing information and a suggestion (with some theoretical examples) is given in Table 6.7.

TABLE 6.7 **A possible matrix for measuring interactions with other habitats**

Interaction	Ecological implications	Social implications	Trends
Mangrove habitat	• Important fish-breeding habitat • Buffer against tropical storms	• Local fisheries dependent on fish • Charcoal from mangroves	Loss due to fish farms
Peat land	• Important habitat • Rare species associated		Afforestation on existing peat Use of peat as fertilizer in plantations
Montane areas	• Existence of rare treeline forests	Site of particular non-timber forest products	No change at present

SOURCES OF FURTHER INFORMATION

In addition to sources about particular impacts, the approaches and techniques of environmental impact assessment can be useful here and many standard methods exist (for example Dalal-Clayton and Sadler, 2005).

Environmental benefits component 4: Influence on climate

TABLE 6.8 **Environmental benefits component 4: Influence on climate**

Methods of data collection	Usually standardized method of measurement of carbon sequestration
Expertise needed	Understanding of methodology
Likely costs	Can be quite high for a thorough study
Pros and cons	This only tells part of the story – more complete or sophisticated analyses of the likely impacts of forests on climate change will be more complex and more difficult

BACKGROUND

Trees play an important role in local climatic patterns through the transpiration cycle, by modifying temperature extremes and by protecting against wind and snow effects. The role that forests play in soil conservation and hydrological cycles also has impacts on local and regional climate – for example by maintaining rainfall patterns and relative humidity.

In addition, forests are important with respect to global climatic patterns through their role in sequestering and storing carbon, which can help to mitigate the potential effects of climate change. Until recently it was believed that young and vigorously growing forests offered the best options for sequestration but many short-term uses such as paper result in carbon being released again a few years later (Brown, 1998). Researchers now believe that old-growth forests also act as important stores, both in woody material and in the humus layer. For example, measurements in undisturbed rainforest in the Amazon suggest that the ecosystem is a net absorber of carbon dioxide (Grace et al, 1995). The standing and lying deadwood in natural forests is also an important carbon store. Deadwood itself releases carbon into the atmosphere during microbial respiration from decomposer organisms, but in ecosystems in cool climates decomposition is very slow so that deadwood acts as a long-term storage site. Much of the carbon in long-lived and slow decaying trees, such as Scots pine, can remain sequestered for over 1000 years. In British Columbia, in forests with a rotation age of 80 years, regenerating stands stored approximately half the wood carbon of nearby old-growth forests (predominant age 500 years), indicating that conversion of old-growth forests to younger managed forests results in a significant net release of carbon (Janisch and Harmon, 2002). Calculations in France suggest that creation of new protected areas (with no logging) can store the same amount of carbon as afforestation (Vallauri et al, 2003). The importance of forests is related to the *total* carbon stored in living and dead trees, in the soil under forests and in wood products.

MEASURING IMPACTS ON CLIMATE CHANGE – PROS AND CONS

The science of measuring such values has improved dramatically over the past decade, in part due to the pressure brought to bear by the UN Framework Convention on Climate Change. From the previous discussion it follows that measuring the impacts of the forest landscape on climate at a global level should include both assessment of carbon sequestration in the landscape and also some indication of how long any carbon that is captured is likely to *remain* sequestered. The latter point will almost certainly have to be approximate but for example, broad figures are available for how long carbon stays sequestered if timber is made into paper or pulp products, packaging and such like.

TABLE 6.9 **Some indicators of climatic interactions**

Indicator	Type of measurement	Units and methodology
Climate stabilization	Global climate change	Carbon sequestration
		Amount of timber removed from the landscape each year and indication of use (firewood, charcoal, pulp etc.)

Forests may have other benefits, including acting as buffers against some of the impacts of climate change such as sea-level rise, extreme weather events and shifting temperature patterns. All of these could be included in the consideration, and in some cases methods for calculating impacts with more precision are also becoming available.

SOURCES OF FURTHER INFORMATION

Standardized methods of measuring carbon are already available (for example Brown, 2002). Attempts to look at the mitigating effect of natural vegetation are less well developed but have been discussed in a preliminary fashion in Dudley and Stolton (2003b).

Conclusions

Environmental benefits are well recognized and relatively easy to measure, at least to the level of accuracy required in a landscape assessment of forest quality. At a landscape scale, larger policy decisions can also be indicative. For example, many governments specifically classify a proportion of their forest estate for watershed protection, which itself provides a useful indicator. Considerable progress has also been made with respect to how environmental benefits could be supported, for example through compensation mechanisms to communities managing forests for wider environmental benefits. However, the widespread adoption of such approaches has yet to be implemented.

7 Social and Economic Benefits of Forest Quality

Latvia is one of the only countries in Europe where real forest edges still regularly occur. In most places demands for land and intensive management systems mean that forests start and finish in abrupt lines, often with a fence to drive the point home. The unique ecology associated with the borderlands is forced back into a few nature reserves or to semi-habitats like hedgerows or urban wasteland. But in the Baltic States the unique mixture of forestry and farming has allowed at least some of the woodlands to spread out gently, tapering away into meadow. It is late summer and in this forest edge one elderly couple are assiduously collecting berries: raspberries that come when trees are felled but also many heath species including cloudberries, which fetch a good price in the markets of Riga. The photographer with us goes over and asks through sign language if they would mind him taking a photograph of them. Berry picking looks like a leisure activity but it is quite big business in a poor country; people from Poland and the Baltics now sometimes travel into northern Scandinavia to collect berries for sale.

Most of the world's forests have been changed to some extent as a result of human activity and many are now effectively cultural landscapes. Forests therefore do not only serve ecological and environmental functions, but are also of more direct economic and social importance to human societies – the third criterion of forest quality. Until recently, most attempts to look at forest condition from a socio-economic perspective focused on one or two key uses, particularly extraction of wood for timber and pulp. Increasing recognition of the wider roles of forests means that assessments have to become more sophisticated and include a far more diverse set of values. Measuring these presents a challenge because they include some elements that are hard to quantify. Below some key social and economic indicators are described and their measurement discussed. These range from commercial considerations through to spiritual and aesthetic values. These include both more-or-less concrete *elements*, such as the amount of timber extracted from the forest, the *processes* that impact upon the forest, such as the type of legal institutions that are in place, and other indicators:

- wood products including timber, pulp and fuelwood;
- non-wood goods and services (NWGS) and non-timber forest products (NTFPs);
- employment, indirect employment and subsistence activities;
- recreation;
- homeland for people;
- educational value including the role of the forests in research;
- aesthetic and cultural values;
- spiritual and religious significance;
- local distinctiveness and cultural values;
- institutions and influential groups
- rights and legal issues;
- knowledge;
- management policies.
- nature of incentives;

Each of these is examined in the following section, with a brief description of why they are important and some initial notes about how they might be measured, with examples where appropriate.

Social and economic component 1: Wood products – timber, pulp and fuelwood

TABLE 7.1 **Social and economic component 1: Wood products – timber, pulp and fuelwood**

Methods of data collection	Usually timber statistics, company records etc.
Expertise needed	No special expertise
Likely costs	Low
Pros and cons	There is a danger of only looking at official statistics in places where there is a large black or illegal economy

BACKGROUND

The use and the sale of wood in its various forms provide the driving force for much of the world's forest management. Uses vary from satisfying local needs, through for example fuel and building materials, to providing resources for global industries such as pulp and papermaking. For approximately half the world's population, wood is still the primary energy source and management for fuelwood can have important implications for other forest values. In the 1970s and early 1980s, shortage of fuel was considered to be a key factor in deforestation but since then opinions have changed; in many areas fuelwood collection is in balance with the forest or, if anything, contributes to degradation more often than outright deforestation (Leach and Mearns, 1988). Exceptions occur when commercial collection and charcoal burning occurs around large cities (Utting, 1991). Use or over-use of fuelwood can result in the removal of several important microhabitats, including dead timber, thus affecting the authenticity of the remaining forest. Today, the main objective of much forest management is commercial timber production, often to the virtual exclusion of other uses. The global economic contribution of forest products, predominantly timber, reached approximately US$400,000 million in the 1990s (FAO, 1997). Uses are gradually changing, generally away from timber and towards fibre and pulp. In Europe during the 20th century there was a gradual shift from using timber predominantly as building material and fuelwood, to today where over half goes to pulp (Dudley, Jeanrenaud and Sullivan, 1996).

Quality of timber from a manufacturing perspective is therefore also a significant aspect of overall forest quality in many areas. This in turn has a profound effect on forest economics, management techniques, and the choice of species and length of rotation cycle.

MEASURING WOOD PRODUCTS – PROS AND CONS

Measuring timber use is one of the best-developed forms of forest assessment, with regular national statistics coordinated by the United Nations. However, data collected at a national level may not translate well to a landscape scale, particularly if some of the timber extraction is unofficial, illegal or for domestic purposes, or if the landscape being assessed does not correspond well with the political boundaries used in collecting data on timber production.

TABLE 7.2 **Some indicators of wood products**

Indicator	Type of measurement	Units and methodology
Wood products	Amount of timber and other wood products	Volume produced per year
	Types of timber	Species, age classes, uses etc.
	Value	Monetary value Non-monetary value

SOURCES OF FURTHER INFORMATION

Statistics from local government bodies and from industry may have to be used in conjunction with other more informal methods, such as interviews and observation.

Social and economic component 2: Non-wood goods and services and non-timber forest products

TABLE 7.3 **Social and economic component 2: Non-wood goods and services and non-timber forest products**

Methods of data collection	Sometimes government statistics, more usually interviews, field research etc.
Expertise needed	Understanding of social surveys
Likely costs	Medium – field visits and interviews
Pros and cons	Because much of this will be informal, hard to get a clear picture – people collecting non-timber forest products may underestimate collection if they feel their activities could be threatened

BACKGROUND

Non-wood goods and services (NWGS) can, from a social and sometimes even an economic perspective, be of greater value than the timber itself, but their importance has tended to be underestimated (FAO, 1995). They include non-timber forest products (NTFPs), food (nuts, fruit, mushrooms, herbs, game and so on), medicines, fodder, building materials, honey, rattan, bamboo, gums, aromatics, ornamental plants and resins (Ruiz Pérez, and Arnold, 1996), along with other services such as grazing in forests and a range of environmental services that have already been described above. This particular indicator of forest quality refers mainly to NTFPs.

Collection of NTFPs is often important from a local economic or cultural perspective, and fodder from forest vegetation is used in many tropical and subtropical areas. Although best known in developing countries, NTFPs are significant outputs from forests all over the world. In developing countries they often play a key role in both subsistence and rural economies and provide both subsistence and cash for the poorest members of society. In richer countries, following a decline earlier in the century, interest is currently increasing again, both as a result of changing fashions and increased leisure activity, and because rising unemployment is forcing people to look at new ways of making money. In the richest countries, immigrant and indigenous communities are often those who collect NTFPs.

Grazing livestock in forests was at one time practised throughout Europe and North America and remains significant in some areas. Grazing includes both use of grass and of seasonal foods such as nuts and fruit. In other countries fodder is collected from forest areas. Trees also provide important shade areas for livestock in hotter countries.

NFTPs remain important in many of the richer countries as well. In Sweden, it is estimated that 100 million litres of berries are collected from the forest every year, along with 20 million litres of mushrooms (FAO, 1986). Collection of medicinal herbs remains significant in Belgium, Greece, Hungary, Italy and Spain, and forest pasture is important in Austria, Bulgaria, Greece, Italy, and Spain (UNECE and FAO, 1994). Between 70–90 per cent of the animal protein consumed in forested parts of Africa comes from 'bush meat' (Sayer and Ruiz Pérez, 1994). Encouragement of traditional NTFPs is seen as a key conservation strategy in parts of the European Mediterranean (Moussouris and Regatto, 1999).

The relationship between people, NTFPs and forests changes with the degree of importance that NTFPs have to lifestyles and economy. Traditional and indigenous users tend to be characterized by a very close and sometimes also a philosophical and spiritual relationship with the forest. Use tends to be sustainable, either because there is a traditional understanding of management or because population density is low enough to avoid over-exploitation. (This situation can change if the motivation for collection changes from subsistence to trade.) In these cases, use tends to be extremely varied, in part because of lack of alternatives.

The social importance and the volume and variety of NTFPs utilized varies widely with the social and economic status of a country, region or people. Collection of wild foods is generally practised most by those in the lower socio-economic classes, including especially indigenous peoples, recent migrants and the unemployed. Some different relationships are illustrated in Table 7.4.

TABLE 7.4 **Relationship between importance of non-timber forest products and impact on the forest**

Importance of NTFPs	Degree of specialization	Impact on forest ecosystems
Necessity: indigenous communities reliant on forest products	Use of a wide variety of products: often many uses for a single species	Can be low impact in terms of species loss, although there are exceptions and it can sometimes profoundly change ecology over time
Full-time or major source of employment	Usually specialized to a limited number of valuable products, either collected from the wild or cultivated	Generally low impact although can sometimes be dramatic – for example when natural forests are replaced by 'plantations' of valuable species or when bushmeat hunting becomes an important means of generating income in an area
Seasonal employment	Usually specialized to a small number of valuable products	Over-collection can sometimes be a problem, for example when migrant workers use an area for a short time
Recreational employment	Usually specialized to a very few products, for example berries, mushrooms	Usually low impact but can sometimes result in over-collection due to poor understanding of ecology

MEASURING NON-TIMBER FOREST PRODUCTS

There have been many attempts to describe values for NTFPs. For tropical forests, an average figure of US$50 per hectare per year was suggested in 1993 although this will at best be subject to massive local differences (Godoy et al, 1993). Export values for NTFPs in Southeast Asia in the late 1980s were calculated at US$32 million for Thailand, US$238 million from Indonesia and at least US$11 million for Malaysia (de Beer and McDermott, 1996). Export values from Peru were more than US$3 million at the beginning of the 1980s, and were over US$42 million from Brazil in 1989, with Brazil nuts and palm hearts being the most valuable components of trade in the latter case (Broekhoven, 1996). In Italy, the total value of non-timber products from forests was estimated at US$75 million in 1986, 45 per cent from chestnuts (Florio, 1992). In practice, economic valuation is only a small part of the real importance of NTFPs because a large proportion of those making use of them will be on the edge of or beyond the cash economy. From our perspective, economic assessments are at best limited because they omit the subsistence values of NTFPs, and these can be amongst the most important. The way of recording NTFPs will vary with location and data availability: measurement can be for example by volume, economic value or other measures such as collection time. Some of these measures are summarized in Table 7.5.

TABLE 7.5 **Some indicators of non-timber forest products**

Indicator	Type of measurement	Units and methodology
Non-timber forest products	Type	List of species used, classified into different uses (herbs, medicines, game, materials etc.)
	Volume	Quantities harvested/year
	Value	Monetary and non-monetary value
	Grazing	Area used for grazing
	Effort to collect	Collection time

SOURCES OF FURTHER INFORMATION

There have been numerous attempts at economic valuation or social valuation of NTFPs and the FAO has produced a thorough overview (Wong et al, 2001).

Social and economic component 3: Employment, indirect employment and subsistence activities

BACKGROUND

The forest provides paid work and subsistence occupation for millions of people around the world. This relationship is intimately bound up with other forest functions.

Paid employment includes both direct employment in the forest and the creation of secondary labour opportunities through support workers and service industries. It is estimated that forestry

TABLE 7.6 **Social and economic component 3: Employment, indirect employment and subsistence activities**

Methods of data collection	Government statistics, company statistics, interviews
Expertise needed	Possibly interview techniques
Likely costs	Low for official employment statistics, higher if informal employment is considered in places where this is important
Pros and cons	Hard to get accurate statistics for informal employment

provides annual subsistence and wage employment equivalent to some 60 million work-years around the world, of which four fifths is in developing countries.

Forest management activities can bring new work into an area and help to maintain traditional skills. Alternatively, forestry operations can also result in an influx of people into an area, creation of temporary work and increased social tensions and environmental impacts. Throughout the developed world, technical innovations and mechanization mean that the number of people employed directly in forest management is often falling rapidly – in one forest area in northern Sweden paid employment has fallen from 127 people to just 6 over a period of 20 years (P. Angelstam, personal communication). To some extent, this labour market has been transferred to other employment linked to forests – particularly the various holiday and recreational areas – although skills are not necessarily transferable.

In addition, the forest supplies employment through more general use of the land under commons agreements, grazing rights, traditional tenure and collection of NTFPs and fuelwood. Forests also provide growing employment in the leisure and tourism industries. Some of the ways in which forests supply employment are outlined in Table 7.7. Management increasingly results in choices, both conscious and unconscious, between different employment options. Studies, for example in Cameroon, suggest that forests can offer more employment as a standing resource than they do by being felled (Ruitenbeek, 1990). High quality forestry should be able to make a positive contribution to community development and to the skills and business opportunities of individuals.

TABLE 7.7 **Types and characteristics of employment in forests**

Type of employment	Characteristics
'Lifestyle'	Immersion in the forest and use of a wide variety of products, typical of indigenous communities
Labour-intensive employment	Aimed either at timber production or collection of specific NTFPs. Relatively large number of low-skilled, low status jobs
High technology employment	Usually involved in wood production. Progressively more technical and highly skilled, with fewer people employed per forest area
Part-time employment	Variable, usually seasonal work often within the 'black economy'. Also tends to vary with broader economic conditions
Indirect employment	Service industries, infrastructure etc. Usually most valuable under conditions of labour-intensive employment

MEASURING EMPLOYMENT – PROS AND CONS

Although direct employment is relatively easy to measure, estimation of indirect benefits poses a more difficult problem, particularly when they themselves depend partly on forest quality, as in the case of recreational use. Statistics for direct employment will be available for many landscapes. Estimating indirect, part-time and 'black economy' or illegal employment is more difficult and will normally have to rely on interviews and estimations.

TABLE 7.8 **Some indicators of employment**

Indicator	Type of measurement	Units and methodology
Employment	People employed	Number of people employed – directly in the forest – indirectly through associated activities Periods of employment
	Value	Monetary value of employment
	Subsistence activities	Number of people using the forest: – for all their livelihood – for part of their livelihood
	Social values	Proportion of local people employed Working conditions (ILO conditions met etc.)

Note: ILO, International Labour Organization.

SOURCES OF FURTHER INFORMATION

Most national or international statistics are too general for this purpose, although they may sometimes break down figures into regional areas. Much of the best information will probably come from interviews with local businesses and communities – forest workers usually have a good idea of how many people are doing similar jobs in the community.

Social and economic component 4: Recreation

TABLE 7.9 **Social and economic component 4: Recreation**

Methods of data collection	Usually data although many long-term monitoring systems exist
Expertise needed	Basic research skills
Likely costs	Low in terms of accessing data, much higher for monitoring
Pros and cons	Still much debate about some of the approaches to measuring visitor value in economic terms

BACKGROUND

Recreational needs are an important facet of forest services in almost all developed countries and, to an increasing extent, also in many developing countries. Activities include walking, picnics, hunting, nature study, food collection, camping and outdoor sports and involve both local people and those from further away. In the UK, research in the 1990s found that some 50 million day visits were being made to forests managed by the state Forestry Commission every year (Department of Environment, 1994). Such a level of recreational use also brings high economic benefits. The Coed y Brennin mountain bike routes in Wales are estimated to bring GB£5 million into neighbouring communities each year.

There are potential and actual clashes between recreation and other forest uses, including ecological services, and also *between* different recreational activities, such as for example walking, bird watching, shooting, use of off-road vehicles and mountain biking. The designation of certain forest areas for particular uses is therefore sometimes essential. Some examples of types of recreational activity and their implications are given in Table 7.10. Accommodating different activities in forests requires regional planning and some compromise. There may be broad differences of opinion between public perceptions of forest uses and quality and those of private woodland owners, and between the requirements of local people and of visitors. Different recreational activities also have different 'carrying capacities' for a forest. For example, a wilderness area has a very low human carrying capacity or it no longer retains the status of wilderness, whereas a leisure park can have a very high carrying capacity.

TABLE 7.10 **Types and characteristics of recreational activities in forest areas**

Type of recreation	Characteristics and requirements
Daily recreation	Use near to home. Often artificial or semi-natural forests (parks, community woodlands etc.) for walking, exercising dogs, jogging etc.
Weekend recreation	Within a few hours of home. Use for walking, picnics, bird watching etc. May use small woodland fragments (sometimes as part of a walk through other landscapes) or artificial woodlands
Specialized recreation	Sports, hunting, nature study etc. Some woods are dedicated to a particular activity for some or all of the time (for example shoots, nature reserves) while other activities can co-exist with general recreation (for example cycling, orienteering)
Mass tourism	Generally use a few well-known forests or woodlands, either with particular historical or biological associations or as a result of specialized tourist development. High level of infrastructure (car parks, cafés, souvenir shops etc.)
Outdoor holidays: family	Use larger areas of woodland, often in popular holiday areas. Use tends to be confined to relatively small areas, for example if visiting a large national park, family use tends to be concentrated on a few areas near to trails, campsites etc. Relatively high contribution to local economy
Wilderness holidays: backpacking	Focus on relatively remote areas, with or without infrastructure (huts, trails). Contribution to the local economy often limited, but may be proportionately more important in areas with low human population and/or low incomes

MEASURING RECREATIONAL ACTIVITY – PROS AND CONS

Techniques in measuring the value and the importance of recreational activities have increased markedly in accuracy over the last few years, although most of these are survey methods over a long period of time and will be less useful for rapid assessments. Many tourist authorities, state forestry and protected area agencies and even private forest management companies compile detailed information about the number of type of visitors and the economic value of tourism. From the perspective of making management decisions, visitor aims are important (for example what proportion of visitors include a national park on their itinerary) along with estimations of how much value they bring to local communities.

SOURCES OF FURTHER INFORMATION

Much of the work on methodologies for assessing visitor numbers and values is associated with protected areas, although the approach is the same for any natural area and good reviews are available from Hornback and Eagles (1999) and Horneman et al (2002).

TABLE 7.11 **Some indicators of recreational activities**

Indicator	Type of measurement	Units and methodology
Recreational value	Number and type of visitors	Number of visitors Hours spent on recreation Proportion of local people amongst visitors Types of activity (walking, hunting)
	Value of recreation to local community	Propensity to pay for recreation Estimation of value of tourism/ecotourism
	Infrastructure	Number of type of tourist facilities

Social and economic component 5: Homeland for people

TABLE 7.12 **Social and economic component 5: Homeland for people**

Methods of data collection	Government and other statistics, field research, interviews
Expertise needed	Social sciences
Likely costs	Can be high for detailed landscape-level surveys but the critical point here is to identify stakeholder groups
Pros and cons	Very sensitive issue, in some countries a large proportion of the forest land may be under claims, and sometimes conflicting claims, and simplified reporting can cause political problems

BACKGROUND

Virtually every forest on the planet has associated human groups, unless these have already been exterminated or driven away. Some indigenous groups have inhabited particular forests for millennia and have developed a sophisticated knowledge of the biology and ecology. Indigenous peoples are amongst the most profoundly affected by loss of forest quality (IAITPTF and IWGIA, undated). Many tropical indigenous peoples first experience contact with Western society when forests are felled, and suffer persecution, introduction of diseases and destruction of their homelands and resources as a result. Throughout the world, communities are in conflict with loggers and other developers. Indigenous people have sometimes suffered from badly designed conservation projects (Pimbert and Pretty, 1995).

The rights and interests of non-indigenous people are also important. Forests provide a range of goods and services that are, to a greater or lesser extent, essential to the survival of human communities. Many of these benefits are lost as forests are destroyed. The FAO estimates that the livelihoods and cultures of 300 million people are closely dependent on the integrity of tropical forests alone (UNESC, 1994). In other countries, forests provide homes through choice because of their aesthetic or landscape qualities, and people living in forest regions also have an important stake in the use and management of the forest. Questions relate to both numbers of people and their livelihood security, land tenure, access to resources and other aspects of well-being. Indigenous communities are associated with most of the world's large remaining forests areas, both temperate and tropical (the main exceptions being places like Tasmania and southern Argentina where past genocides or disease epidemicss have destroyed indigenous populations).

MEASURING THE IMPORTANCE OF FORESTS AS HOMELAND – PROS AND CONS

Most ways of measuring relevance of forest as homeland focus on the numbers of people and their rights and access to land and resources. From the perspective of a forest quality assessment, the critical point is to make sure that all the different stakeholder groups are identified and if possible their homelands mapped, at least approximately. (Precise mapping is often difficult and may be irrelevant in the case of traditional lands, where ownership patterns can change over time and may not be the same with respect to all resources.) In some situations this includes people who use the forest on a temporary basis – for instance as part of a regular nomadic cycle or as an emergency resource in times of drought or other difficult conditions.

SOURCES OF FURTHER INFORMATION

A range of methodologies exist for identifying different levels of wellbeing and rights (Colfer et al, 2000a; 2000b). More recent research also looks at options for examining the relationship between local communities and their forests, including the sustainability of community managed forest landscapes (Warner, 1995; Ritchie et al, 2000).

TABLE 7.13 **Some indicators of homeland**

Indicator	Type of measurement	Units and methodology
Homeland	Number and type of people living in the forest	Number of people Number of settlement, homes (first/second) Proportion indigenous/non-indigenous
	Land tenure issues	Types and existence of land rights
	Stability	Number of people settled for over 20 years Rate of in–out migration

Social and economic component 6: Historical importance

TABLE 7.14 **Social and economic component 6: Historical importance**

Methods of data collection	Existing maps and data, if necessary basic field research
Expertise needed	Archaeologists if field research is needed – in many cases information will already be available
Likely costs	Quite high in the case of surveys – low elsewhere
Pros and cons	Need to look at both built environment and also historical forest management systems

BACKGROUND

Many important historical sites now exist in and are to some extent protected by, dense forest. Some of these forests are referred to as 'natural' although they must actually be secondary – for example the ancient Roman city of Termessos in Turkey is in forest that is now a nature reserve because of its high biological value, and the ruins of Angkor Watt in Cambodia are surrounded by dense 'natural' forest. Other examples include the Mayan civilizations of Central America, many old cultures in Southeast Asia and Roman remains in Mediterranean Europe.

In addition to the presence of historical sites in forests, some forest management systems have their own historical and cultural significance. For example, the medieval oak (*Quercus robur*) and hornbeam (*Carpinus betulus*) coppice system in England is still carried out today to preserve an old, albeit non-natural, forest type (Kirby, 1992b). In the Mediterranean countries, ancient olive groves now have considerable historical and cultural importance. In Sumatra, 'forest gardens' have been cultivated for hundreds of years. To the outsider, they appear to be identical to the natural forests but produce a sustainable crop of timber and fruit. They now have both an historical and a biological value (Nugroho and Siliew, 1997). On the savannah edge in parts of Africa, human communities deliberately create and protect woodlands, creating cultural landscapes that would quickly disappear without continued intervention (Fairhead and Leach, 1996).

Although in many countries historical artefacts will have official recognition and protection, in other places their identification and protection lies more with traditional users or non-governmental organizations (NGOs).

MEASURING HISTORICAL IMPORTANCE

In developed countries, maps, records and government and non-governmental agencies will quickly be able to identify known sites, although there is scarcely anywhere in the world where such surveys are complete. In less developed countries or less well surveyed areas, some basic survey work may be required, often using aerial photography, and interviews with local people can also be a quick way of locating sites.

TABLE 7.15 **Some indicators of historical importance**

Indicator	Type of measurement	Units and methodology
Historical/cultural value	Cultural artefacts	Number, type and location of cultural artefacts: – officially protected sites; – World Heritage Sites; – other important sites
	Cultural landscapes	Presence and type of culturally important landscapes within the forest, for example: – types of forest management; – grazing regimes etc; – reference in literature, folklore

SOURCES OF FURTHER INFORMATION

There are several useful sources on methodologies to study forest history, such as Agnoletti and Anderson (2000). Local museums can often help and in many cases local knowledge remains important.

Social and economic component 7: Educational value, including the role of forests in scientific research

TABLE 7.16 **Social and economic component 7: Educational value, including the role of forests in scientific research**

Methods of data collection	Interviews, contact with local tourist companies and research organizations, web searches
Expertise needed	Low
Likely costs	Low
Pros and cons	Need to include formal and informal education and also formal and informal (for example amateur) research

BACKGROUND

Forests perform an important educational function in many cultures. This includes practical education in terms of learning how to live in and use the forest, formal education for school and further education students and education in a broader sense of providing information and experience to the general public. Key elements include the provision of information in printed form and increasingly also on the web, through visitors centres, guided trails and the use of personal guides. While most effort focuses on protected areas and increasing number of commercially-managed forests, state forests also devote resources to educational purposes. Forests are living laboratories in which scientists can learn more about ecology and biology. A proportion of natural and managed forests therefore needs to be maintained as 'reference forests' for the purposes of scientific research. For example, in Europe the European Commission is currently coordinating attempts to set up a network of strict forest reserves for research purposes (Parvianen et al, 2000). Similar examples exist around the world. For example, the H J Andrews Experimental Forest was protected by the US Forest Service in 1948 as part of a network of forests intended to serve for studies by the Service's scientific research branch. It is administered cooperatively by the United States Department of Agriculture Forest Service Pacific Northwest Research Station, Oregon State University and the Willamette National Forest. Long-term field experiments have focused on climate dynamics, stream-flow, water quality and vegetation succession; over 3000 scientific publications have used data from the forest (Luoma, 1999).

MEASURING EDUCATION VALUE

The indicator covers two main issues: the *potential* importance of the landscape for research and education and its *actual* use for these purposes. The first is by its nature likely to be a qualitative judgement, although the presence of endemic species, unique ecosystems or other non-replicable attributes provides some quantifiable indicators. The second, actual use, is possible to measure in terms of visitation and money spent.

TABLE 7.17 **Some indicators of educational value**

Indicator	Type of measurement	Units and methodology
Educational value	Importance	Uniqueness of particular sites
	Current use	Number of educational visits Number/use of educational infrastructure, for example: – guided trails; – test sites for scientific research
		Web searches of the scientific literature for reference to particular sites

Social and economic component 8: Aesthetic and cultural values

TABLE 7.18 **Social and economic component 8: Aesthetic and cultural values**

Methods of data collection	Interviews, existing literature
Expertise needed	Some knowledge of social sciences research skills useful
Likely costs	Low
Pros and cons	As with homeland, cultural values are hard to grasp and sensitive; cultural values may be completely different between different stakeholder groups – indeed may need to be reconciled

BACKGROUND

Forests often touch on the deepest levels of emotional and cultural feelings in humans, and have inspired artists and philosophers for centuries. They feature in folklore, novels, poetry and the visual arts and often act as inspiration for musical composition. A forest's attractiveness is of key importance from its recreational perspective and for the people living nearby. Aesthetic values are notoriously difficult to measure, although many attempts have been made even using strictly economic criteria (Magill and Schwartz, 1989) and theoretical research frameworks exist (Bourassa, 1991). Factors recognized as having an impact on aesthetic perceptions of the forest in Europe include age and diversity of trees, the proximity of other wildlife, quality of light, quietness, forest mosaic and often the presence of water, but these qualities are likely to change between cultures, socio-economic background, age and gradual historical shifts of perspective. Issues of 'wilderness' create particularly strong feelings (Martin, 1982), with opinions ranging from attraction for wilderness values amongst many urban dwellers, to more traditional reactions against wilderness as a dangerous or unproductive area amongst traditional rural communities. Many indigenous peoples view the concept of 'wilderness' as insulting in that many places labelled as wilderness areas have been inhabited and managed, albeit subtly, by indigenous peoples for many hundreds or thousands of years. Such debates are made more difficult because 'wilderness' itself is a social construct that differs in interpretation around the world (Shepheard, 1997), with some countries even having legally defined wilderness values. Reconciling such differences in terms of management decisions sometimes requires major trade-off negotiations.

There is no simple relationship between aesthetic values and naturalness. Although many natural forests are judged to be beautiful, so are formal forest gardens and arboreta (Mitchell, 2001). Historical associations and education inform aesthetic values. For example, understanding of ecological needs can change perceptions towards 'tidiness' of managed forests or composition of species. Greater understanding about historical forest loss in upland Britain has changed attitudes towards restoration of natural forests; previously the bare moors were considered to be 'natural'. Many forests that are, to an ecologist, clearly managed landscapes appear to the casual onlooker to be natural areas, although with greater understanding about ecology these misperceptions are gradually changing.

MEASURING AESTHETIC VALUES – PROS AND CONS

From the foregoing discussion it is clear that surrogate indicators are particularly important here. In landscapes where some official recognition of the aesthetic value exists, this can suffice, although in many areas judgements will have to be made either through interviews or by using recognized measures of visual impact, although the last is fraught with the problems of interpretation and agreement. It should also be remembered that what may be 'obviously' a beautiful view to one set of onlookers may have no aesthetic quality to another.

SOURCES OF FURTHER INFORMATION

Expert systems to evaluate landscape quality exist (for example Buhyoff and Miller, 1998), although simpler approaches will usually be adequate.

TABLE 7.19 **Some indicators of aesthetic and cultural values**

Indicator	Type of measurement	Units and methodology
Aesthetic value	Official recognition of aesthetic value	Designated Areas of Outstanding Natural Beauty or similar
	Other measures	Visual impact, for example: – area to edge length; – presence of water; – mixed ages and species of trees; – inspiration for artists Interviews with local people and visitors

Social and economic component 9: Spiritual and religious significance

TABLE 7.20 **Social and economic component 9: Spiritual and religious significance**

Methods of data collection	Official records, interviews
Expertise needed	Some social science research skills are useful
Likely costs	Interviews, field visits
Pros and cons	In some cultures, part of the sacredness may be the site's secrecy, so identifying these in a survey is difficult

BACKGROUND

All major religious philosophies acknowledge the importance of forests and other natural ecosystems. Some cultures and religions, such as the Buddhist and Hindu faiths, have especially strong links with forests, as do many indigenous groups (Dudley et al, 2006). Significance can be attached to particular trees, forest glades ('sacred groves') or whole regions (Porteous, 1928). Some examples are given below (shortened from a table in Dudley et al, 2006; other sources are Pakenham, 1996; 2002). Spiritual considerations can be an important motive for managing and preserving forests (Lee and Schaff, 2003). Sacred trees are found all over the world, including the baobab in East Africa, sacred prayer trees in Estonia and the yew (*Taxus baccata*) planted in English churchyards (Hartzell, 1991). Religious centres provide refuges for primary woodlands or particularly old trees. Whole forests can also have significance. Coastal *kaya* forests in Kenya have important spiritual significance and are managed by the National Museum Service (Negussie, 1997). In Xishuangbanna, China, Buddhist 'white elephant temples' guard the sacred groves of the 'dragon hills' (Sochaczewski, 1999). Such values are seldom overtly incorporated into forest planning at present. However, they are beginning to be recognized, for example by a five-year UNESCO Sacred Sites and Nature Conservation programme. Tree worship is important in some areas and the act of planting a sacred tree can also be important. Rituals are conducted in forests and some religious institutions themselves manage forests, although most of these are small (Ingles, 1995).

TABLE 7.21 **Some examples of sacred trees**

Country	Species	Details
China, Korea and Japan	Gingko (*Gingko biloba*)	Revered by Buddhists in China and Korea and long planted as a temple tree. The species was introduced into Japan less than 1000 years ago. A gingko tree survived 800 metres from the epicentre of the nuclear bomb at Hiroshima
India, Nepal, Sri Lanka etc.	Banyan tree (*Ficus benghalensis*)	Protected because it is considered sacred but also because of the shade it provides. Strict regulations control its use; some are reserved for elephant fodder. Some scholars propose that the tree of life in the Garden of Eden was a banyan
	Bodhi tree (*Ficus religiosa*)	The tree beneath which the Buddha gained enlightenment and the most sacred tree to Buddhists, also considered sacred by Hindus
Madagascar	Baobab (*Adansonia*) – six species	Considered holy and to be the home of spirits, often in consequence protected even when the rest of the forest has been logged or cleared
Europe	Yew (*Taxus baccata*)	Sacred to the early Celtic and Nordic tribes, believed to be immortal and a symbol of everlasting life. Some churchyard yews are so old that they predate Christianity
New Zealand	Kauri (*Agathis australis*)	Sacred trees to the Maori, who believed that it possessed its own spirit. The oldest known living specimen is 2100 years old
	Totara (*Podocarpus totara*)	Believed to share a common ancestry with the Maori and therefore to be an elder of living Maoris
Chile	Monkey puzzle tree (*Araucaria araucana*)	Sacred to the Pehuenche people who regard it as a 'mother' and believe that god created the trees for them and that it is their duty to protect them
United States	Giant redwood (*Sequoiadendron giganteum*)	American redwoods are sacred to the Tolawa people of the coastal areas of California and Oregon The oldest known specimen is estimated to be 2700 years old

Sacred values are not necessarily attached to the most natural forests, although there is often a link. Some of the few remaining patches of lowland forest in parts of Japan are associated with Shinto temples, giving the trees both a spiritual and a biological importance. Small patches of forest preserved for their spiritual values can be seen in otherwise deforested areas of countries such as Lao PDR. In Borneo, fruit gardens called *tembawangs* are planted and therefore wholly 'unnatural' but in heavily settled areas they serve a dual function of acting as burial grounds and also as a reservoir for biodiversity. Here food production, spiritual importance and biological value go hand-in-hand; indeed in forests the sacred is often associated with the mundane. The Kaya forests of Kenya are remaining fragments of natural forest precisely because they have religious significance although they also supply NTFPs. Spiritual values therefore often have a direct correlation with other values, such as those of biodiversity conservation. Some different levels of spiritual values are summarized in Table 7.22.

TABLE 7.22 **Spiritual values of forests**

Different spiritual values associated with trees and forests				
Small areas				Large areas
Specific				General
Temples (e.g. ancient trees around Shinto temples in Japan)	Particular tree species associated with spiritual values (e.g. yew, baobab)	Sacred groves (e.g. in India, Ghana, Kenya)	Larger areas associated with spiritual values (e.g. association of trees with spiritual values in Hindu religion)	General spiritual feeling (e.g. personal philosophy, woods of value to certain villages or families)
Shrines (such as tree shrines found in Switzerland)		Sacred animals or plants found in forests (e.g. the tiger in some Asian societies)		
Burial sites (e.g. *tembawang* in Indonesia)				
Sacred trees (e.g. in Estonia)				

MEASURING SACRED VALUES – PROS AND CONS

Measurement of such values must inevitably be qualitative; there is no recognized scientific method for assessing the number of gods in a forest and comparing their importance. But identification of such values as sacred groves, sacred trees, burial sites and temples can play a critical role in determining land-use patterns and can often also have a direct relationship to conservation values. While some such values will be easily identified, others will be far more difficult, more localized and even deliberately hidden from outsiders. Time, the building of trust and careful research will be necessary in some communities to identify spiritual values within a forest landscape.

TABLE 7.23 **Some indicators of sacred values**

Indicator	Type of measurement	Units and methodology
Spiritual and religious importance	Existence of spiritual sites	Number and type of sacred groves, trees, burial sites, religious buildings etc.
	Importance of sites	Interviews with local people Propensity to pay for spiritual values Number of visitors

Social and economic component 10: Local distinctiveness and cultural values

TABLE 7.24 **Social and economic component 10: Local distinctiveness and cultural values**

Methods of data collection	Interviews, workshops
Expertise needed	Some knowledge of social science research techniques useful
Likely costs	Low
Pros and cons	Will vary between stakeholder groups

BACKGROUND

An additional complication for planners and forest managers is that these wider issues are seldom assessed equally across the whole community. Each area of forest or woodland is not only unique from an ecological perspective, but also has a unique and often very local set of cultural associations. People living in or near a forest often have different ideas about the importance of aspects of forest quality than those held by visitors or planners. Making efforts to incorporate local cultural and social values into forest management is therefore an important aspect of quality. Priorities for local people frequently include NTFPs, fuelwood, particular historical associations (which may not have any national or heritage significance), accommodation of aesthetic issues relating to local communities and past or current events, and local recreational issues (Clifford and King, 1993). Local values may not necessarily be the same as those of other stakeholders (e.g. industry, conservationalists), and can sometimes be in conflict with requirements for timber production, biodiversity conservation and other uses, but may be amongst the most powerful values held. In many or perhaps most cases, a compromise between local and national or regional needs will have to be reached in any assessment and subsequent planning or implementation of management.

MEASURING LOCAL VALUES – PROS AND CONS

By their nature, such values are not open to generalized or externally-determined indicators. Local values may overlap with other indicators discussed above including spiritual, aesthetic and historical values, but are addressed separately here because they are often missed in broad assessments. Determining local values is therefore *only* possible through direct access to local opinions; this will often have to be directly through interviews but could for example draw on locally produced written sources (newspapers, community magazines, community websites, and local associations).

TABLE 7.25 **Some indicators of local distinctiveness**

Indicator	Type of measurement	Units and methodology
Local distinctiveness	Areas of high value to local people	Identification, description and mapping of areas of high values through: • interviews (which should aim to find out why particular forests are important); • local newsletters and magazines; • community websites; • local organizations and centres of religious worship

Social and economic component 11: Institutions and influential groups

TABLE 7.26 **Social and economic component 11: Institutions and influential groups**

Methods of data collection	Discussions with stakeholders
Expertise needed	Understanding of policy issues
Likely costs	Low
Pros and cons	A critical part of the overall assessment is to find out who has most influence over decisions about land use

BACKGROUND

The forest can be influenced by institutions of all types and sizes, and by those existing in the immediate locality and those that may never have a direct contact with the forest at all. Institutions therefore range from village associations through a range of local and regional bodies (local authorities, religious institutions, companies, NGOs, research groups and so on) all the way to national governments and international bodies. This indicator also relates to governance, and the way in which the landscape is influenced and how decisions are made. From a practical point of view, understanding who has power to shape the landscape, who makes the decisions and if and how they might be influenced are all critical factors in completing and using an assessment. Although we use the term 'groups' here, in some cases the influence may come mainly from one or more powerful individuals.

MEASURING INSTITUTIONS AND INFLUENTIAL GROUPS – PROS AND CONS

This indicator is less about quantitative measurement of the size and importance of institutions within a landscape but more an initial identification of relevant stakeholders that should be involved in any assessment and subsequent planning or negotiation. Measurement therefore involves identification and simple classification although in many cases some indication of the strength, influence and views of such institutions might also be useful, along with the parts of the landscape where particular institutions have influence.

TABLE 7.27 **Some indicators of institutions**

Indicator	Type of measurement	Units and methodology
Institutions	Types and nature of institutions affecting the forest landscape	Listing and assessment of relevant institutions at the following levels: – local; – regional; – national; – international

SOURCES OF FURTHER INFORMATION

This is less a measurable indicator than a note to assessors that good assessments will also have some analysis of power relations. Such analyses should also look specifically at how the least powerful members of society fare in the decision-making process (for example Chambers, 1997).

The remaining indicators are related to *process* functions. Rather than goods and services of various sorts, these indicators refer to legal issues, incentives for management and less tangible issues such as knowledge. Although sometimes difficult to gain an understanding of, these process issues are often critical to the status and quality of the forest landscape and the well-being of people who live there. They will in general involve an understanding of conditions at a regional level, rather than of individual components within a landscape, and thus assessment approaches will be different from those described earlier.

Social and economic component 12: Rights and legal issues

TABLE 7.28 **Social and economic component 12: Rights and legal issues**

Methods of data collection	Discussions, web searches
Expertise needed	Some understanding of law useful
Likely costs	Low
Pros and cons	Need to consider both traditional and formally legal issues

BACKGROUND

The laws surrounding the forest – whether officially recognized or customary rights – have a critical impact on the way in which the forest is used. For people living in the forest, land tenure is often a fundamental issue and this in turn has an impact on the way in which the forest is maintained. Uncertainty of land tenure often leads to degradation and loss of quality because users have no long-term involvement in the land or security over resources (Fisher, 1994). Alternatively, private control without a strong framework of national legislation can sometimes also result in loss of quality when collective values are sacrificed for private gain. Legal controls come at a number of levels, as outlined in Table 7.29.

There is overlap between some of the categories above. Laws include both the rights to forest resources and various forms of access and in many cases control the extent to which private or corporate owners of forests can alter the structure of the forest. Such laws are almost always a balance between the needs and desires of the individual or the local community, and the needs and desires of the wider community or more powerful individual. In many countries, laws are currently changing, often giving greater control back to local communities (Molnar et al, 2004).

TABLE 7.29 **Brief categorization of rights and legal issues**

Level	Details
International	Global regimes (conventions, treaties, international laws, international agreements) and regional agreements (regional criteria and indicator processes etc.) (Tarasofsky, 1996)
National	Laws, regulations, land settlement agreements
Regional	Laws relating to sub-national units (states, provinces, counties, cantons etc.)
Unofficial	Codes of practice, voluntary agreements, certification systems
Traditional	Traditional tenure rights, traditional agreements, village controls

MEASURING LEGAL ISSUES – PROS AND CONS

A thorough grasp of laws and rights over forests is a critical element in an overall understanding of both forest quality and options for managing to maintain or improve quality in the future. As in the case of institutions, this indicator is likely to be concerned mainly with ensuring that an understanding of relevant national and local laws is included in the assessment. Identification of the existence and area of influence of local or traditional rights and agreements is particularly important and in many landscapes may involve direct contact and interviews with the communities themselves. Information is likely to be recorded most effectively through a mixture of description and mapping.

TABLE 7.30 **Some indicators of legal and customary rights**

Indicator	Type of measurement	Units and methodology
Rights and legal issues	Ownership of forests and forest resources	Ownership patterns Security of tenure (type of tenure and disputes) Length of tenure
	Access/use rights	Type and nature of access/use rights: – for physical access (walking, recreation); – for resources (game, firewood); – timber concessions; – mineral rights
	Intellectual property rights	Ownership of knowledge, for example: – biodiversity rights; – uses of forest products

SOURCES OF FURTHER INFORMATION

A number of tools and methodologies exist to understand these including a pyramid of governance developed by the International Institute for Environment and Development (Bass and Mayers, 2001). There are several (rather dated) sources giving guidance on how international policy impacts on forest laws (for example Humphreys, 1996; Tarasofsky, 1999), but little general guidance on national laws, which have to be researched in country. The ECOLEX website (http://www.ecolex.org/index.php) gives access to good information about environmental law. The Programme of Work on Forest Biological Diversity from the Convention on Biological Diversity (CBD) (UNEP, 2004) also provides a useful framework for government commitments on forest conservation issues.

Social and economic component 13: Knowledge

TABLE 7.31 **Social and economic component 13: Knowledge**

Methods of data collection	Interviews, knowledge gained during the course of the assessment
Expertise needed	Basic interview skills
Likely costs	Low
Pros and cons	This is principally about identifying levels of knowledge and gaps in data and understanding

BACKGROUND

This indicator includes the level and type of knowledge about the forest. Knowledge systems range from traditional or indigenous knowledge to scientific or empirical knowledge and include both qualitative and quantitative elements. Ownership of knowledge is itself a critical issue in some cases; the issue of intellectual property rights over resources from the forest (for example biodiversity) is a key question in point.

MEASURING KNOWLEDGE – PROS AND CONS

As with other process orientated indicators, the most important issue here is identifying the types and levels of knowledge (and perhaps from the practical viewpoint of completing the assessments as effectively as possible, identifying alternative sources of information). This indicator will also provide important background information about how accurate data on other indicators are likely to be.

TABLE 7.32 **Some indicators of knowledge**

Indicator	Type of measurement	Units and methodology
Knowledge	Extent of detailed knowledge about the forest	Assessment of types of knowledge, for example: – indigenous; – local; – traditional; – cultural; – scientific research; – business information

This is not a precise survey. In practice, such information can be summarized in a brief matrix to provide an indication of both knowledge levels and their impact on the assessment. A proposed matrix is outlined in Figure 7.1.

Type of knowledge*	Level of knowledge				Access by assessment team	
	High	Medium	Low	Zero	Access	No access
Indigenous						
Local						
Traditional						
Cultural						
Scientific						
Business information						
Other (specify)						

FIGURE 7.1 **Matrix for recording levels of and access to knowledge of forest quality in a landscape**

Social and economic component 14: Management systems

TABLE 7.33 **Social and economic component 14: Management systems**

Methods of data collection	Existing assessments, interviews, possibly tailored assessments
Expertise needed	Understanding of forestry and management
Likely costs	Low if assessments exist, higher if efforts are needed to find out what is happening throughout the landscape
Pros and cons	Both formal and informal management systems should be included in the analysis

BACKGROUND

Choice of management system is also an important indicator, underlying many of the issues raised above. 'Management' can range from deliberate non-intervention (for example in a designated wilderness area) to various intensive forms of use. It also varies from management directed primarily at a single end product (such as timber, hunting or recreation) to management that tries to satisfy many different user groups – multi-purpose forest management.

On a landscape scale, this can, and usually will, also include individual areas that are managed for specific needs, merged together into a more varied patchwork of forest types aiming to fulfil different functions.

MEASURING MANAGEMENT SYSTEMS – PROS AND CONS

Many assessment systems already exist for management. Examples include various forms of forest management certification, for example those of the Forest Stewardship Council (FSC) (Elliott, 1995; Viana et al, 1996) and scorecards for assessing steps towards good forest management. A group of assessment systems is also being developed for protected area management effectiveness, grouped around a framework agreed by the IUCN (The World Conservation Union) World Commission on Protected Areas (Hockings et al, 2000) and with a range of methodologies for different situations and different levels of detail (Kothari et al, 1989; Cifuentes et al, 1999; Ferreira et al, 1999). However, most of these deal at site level and scaling up to landscape may entail summing different assessments or developing indicators suitable for particular situations. Many of these assessment systems also differ in the level to which they allow for participation so that some provide only one particular viewpoint in determining what constitutes 'good' management.

Forest quality assessments can draw on all of these, if they already exist, for part or all of the forest, but there will probably not be the time or resources to put complicated certification or analytical systems in place as a component of the assessment and in this case simpler approaches will be needed. Use of a simple scorecard to assess effective management of different forest management units is one option (see for instance Dudley and Pollard, forthcoming), although this will incur a lot of work. At a landscape scale, broader consideration of management planning, policy framework and opportunities for stakeholder participation may be more useful.

TABLE 7.34 **Some indicators of management**

Indicator	Type of measurement	Units and methodology
Management and land use	Type of management (forest management system, protected area management, community woodlands etc.)	Existence of forest management plans Number and area of certified forests Number, area and type of protected areas Regeneration methods Legal and policy framework Existence of landscape-level planning Surrounding land use
	Opportunities for stakeholder participation	Framework for stakeholder participation: – degree of stakeholder participation; – nature of stakeholder participation; – extent that management responds to local needs/desires

SOURCES OF FURTHER INFORMATION

Some good websites exist for assessment, for example those of the Forest Stewardship Council (http://www.fsc.org) and the PEFC (http://www.pefc.org/internet/html), but at a global scale these systems are still very limited.

Social and economic component 15: Nature of incentives

TABLE 7.35 **Social and economic component 15: Nature of incentives**

Methods of data collection	Official documents, interviews
Expertise needed	Understanding of development issues
Likely costs	Low – this information should emerge gradually during the course of the assessment
Pros and cons	Formal and informal incentives should be considered – the latter may be more difficult to identify

BACKGROUND

Lastly, an important process indicator is the nature of the incentives influencing or driving forest management – economic, legal, traditional, spiritual and emotional. Incentives can act against each other; for example the drive to make money from a forest may conflict with the desire to maintain a beautiful landscape, even within a single person. Developing or identifying incentives to encourage high quality forest management is a critical issue that relates directly to management issues. Following a forest quality assessment, the institution or body involved (local community, protected area manager, timber company and so forth) should have *a concise picture of the strengths and weaknesses in the forest landscape in terms of potential social, environmental and economic benefits.* Hopefully it will also result in a more general understanding of different viewpoints. The initiators are then faced – if they wish – with the challenge of working to maximize the benefits and minimize the problems.

However, most forests do not exist in isolation from human communities; any changes in management approach will entail discussion, compromises and agreements, and a complex web of local and sometimes also of national or international politics. Even if a local community is united in its assessment of what needs to change, people still need the resources to make and sustain these changes. In most countries, forests also have to take their place within a wider landscape that includes other land and water habitats and other uses, further complicating the situation.

High quality management is therefore seldom if ever a free commodity. It needs both psychological and physical incentives and sometimes also a firm regulatory framework (Bass and Hearne, 1997). Making progress towards a higher quality forest estate therefore usually requires the use of a range of different tools and techniques, grouped under the following headings (distinctions are not always precise):

- *regulations* to ensure that minimum environmental and social conditions are met;
- *education* and *training* to provide skills and understanding of the issues;
- *planning* and *assessment* tools to aid implementation of any changes;
- *incentives* and *compensation* mechanisms to encourage and to facilitate best practice.

Regulations
Regulations are essential to ensure that for example powerful minorities do not subvert the wishes of the majority – at both local and national level. National and local laws provide a framework that can, at least in theory, protect against expropriation of communal benefits by a minority and put local decisions within a national or regional context (Humphreys, 1996; Tarasofsky, 1996).

Education and training
Education and training are needed in many cases to help communities to both value and manage forest resources. While some forest-dwelling communities will already have a highly sophisticated understanding of the forest ecology – and in fact can serve as trainers themselves – in other cases communities may have moved to the forest recently, forgotten old skills or need additional help in realizing the potential of new opportunities. The importance of education and training is a constant factor in all the work on forest quality. It is important to recognize that such education can work in two directions because 'experts' often have much to learn from local communities. Educational opportunities may include bringing different groups together or informal teaching alongside more traditional approaches to extension and training.

Assessment systems and guidelines, codes of practice and technical manuals
These all help people charged with managing a forest landscape have information about best practice. Access to databases of such information is an important component of any attempt at sustainable management on a landscape level. Some of the issues that need to be considered range well beyond conventional forest management – such as farming standards and codes for tourism.

Compensation and incentives
Compensation and incentives provide both the encouragement and the means to practise sustainable strategies. They range from basic incentives – such as the desire to maintain natural systems and a pleasant environment – to economic compensation mechanisms and incentives. Local communities will often have to assemble a mixture of economic incentives and compensation mechanisms to pay for sustainable forest management.

MEASURING INCENTIVES – PROS AND CONS

Where an assessment is used for planning rather than for simply producing a status report, identification of possible incentives is a critical first step in determining next steps. Some possible indicators are suggested in Table 7.36.

TABLE 7.36 **Some indicators of incentives**

Indicator	Type of measurement	Units and methodology
Nature of incentives	Nature of incentives for particular types of forest management	Assessment of types of incentives, for example: – social; – political; – economic; – cultural

Conclusions

It is perhaps significant that, even when developed by conservation organizations, the human indicators are more complex than those measuring the ecosystem. To some extent this reflects the fact that people have tended to lengthen the list of indicators referring to their immediate needs during the development of the methodology. However and more fundamentally, this also indicates the real complexities surrounding these issues.

Part 3
Case Studies

So far most of this has been theory. The practice is never quite as easy. There are several dozen excellent guides to participatory approaches published around the world, but few if any perfect examples of participatory approaches in practice. The mess and confusion of real life gets in the way. Alternatively, perfection is seldom required, and there are plenty of examples of efforts to get stakeholders involved in projects that delivered more benefits as a result.

The next part of this manual summarizes some of our experiences in using the forest quality methodology in three continents. There is a great deal of variation in exactly what was needed, and in available resources, skills, data availability and follow-up. Quite a few of the projects are still ongoing and it is too early to report long-term outcomes. It is fair to say that we would not do any of them in exactly the same way again, but at the same time none of them were complete disasters. And because they all approached the issue of forest quality from quite different perspectives, they each tell us something slightly different about how the methodology described in the earlier part of the book might be applied in practice.

Case Study 1
A Participatory Assessment in Wales, UK

A participatory assessment of forest quality was undertaken for a local non-governmental organization (NGO), the Dyfi Eco Valley Partnership (DEVP), in Mid Wales. The aim of the assessment was to help the partnership develop a 'vision' for its sustainable rural development work as it related to forests in the watershed. Wales poses some interesting challenges in terms of rural development. Once a separate country, which retains its own language, Wales is now part of the United Kingdom and although there is quite a strong nationalist movement, this has never attracted majority support for independence. However, there has been strong support for greater local autonomy and a Welsh Assembly now provides greater national direction than has been the case in the past. The forest quality assessment took place at a time when these changes were starting to have real impacts on policy and addressed forests in one of the areas where Welsh culture remains strongest; over 50 per cent of local people still speak Welsh.

In prehistoric times, Wales would have been almost completely forested but there has been a dramatic decline in forest cover throughout the country, from 90 per cent to only 10 per cent. Only a few remote areas, including the Dyfi estuary that was the focus of the research, reach 25 per cent forest cover. However, although the catchment still has a relatively high proportion of tree cover, plantations dominate the area and in the recent past native woodlands have been felled to create space for exotic plantations. Protection is also quite high as half the area is in the Snowdonia National Park, an IUCN (The World Conservation Union) Category V protected landscape and many individual forest patches are within other, stricter forms of reserve.

The valley is the home to various 'green' businesses. Since 1998, the DEVP has been working in the catchment as an NGO fostering sustainable development, with members from private companies, voluntary bodies, individuals, local authorities, the National Park and the Welsh Development Agency. The current project aimed to develop a vision of forest quality to help direct the DEVP's future work.

The *first stakeholder meeting* was held in the winter. Ahead of the meeting, an information sheet in English and Welsh was sent to 500 people and articles appeared in three local papers. The meeting took place at the local school. Representatives came from the two state forest bodies, Forest Enterprise and the Forestry Commission, Tir Coed (a partnership approach to forest restoration), the council, farmers' organizations, the Centre for Alternative Technology, local farmers, landowners, foresters, woodworkers and residents. The meeting introduced the concept and sought approval for the aims, area and key indicators. There was enthusiasm for the assessment. Most of the discussion focused on selection of indicators, which were chosen by brainstorming. The indicators focused on social and economic issues; issues of authenticity were judged important but not developed, perhaps because no conservation organizations were represented. Several people offered information, ideas and contacts.

The *assessment* focused on two major areas, which reflected the main aims of the assessment agreed on at the first stakeholder meeting:

1 Building a vision of forest quality for the Dyfi Valley;
2 Researching the indicators of forest quality identified by the stakeholder meeting.

Note: The area covered by the Dyfi Eco Valley Partnership and therefore the area surveyed by the assessment. Afon Dyfi, the main river in the catchment, is the traditional border between north and south Wales.

FIGURE CS1.1 **The area covered by the Dyfi Eco Valley Partnership**

The basis of vision was drawn from discussions. Many people were visited or contacted by telephone. Initial proposals were circulated at a second stakeholder meeting to stimulate discussion. The first stakeholder meeting had not identified all the indicators that were important to people in the catchment and additional indicators were added during the research phase, following further one-to-one meetings. It became clear that one public meeting did not provide a sufficient enough forum to collect public opinion and in fact the study would have been more participatory if there had been time to meet with particular groups of stakeholders (for example farmers' organizations or the local council).

A SUMMARY OF THE VISION AGREED FOR FORESTS IN THE DYFI CATCHMENT

There is general contentment with the amount of woodland and of protected areas in the watershed, but not with the composition or distribution of trees, and the assessment and accompanying meetings reached broad agreement on the need for:

- more broadleaves in the landscape;
- restoration on some previously wooded sites and perhaps removal of trees from sites that would not naturally be forest covered;
- more benefits for local people;
- a multipurpose forest estate;
- a coordinated approach to grant support for woodland management.

There was a wider range of opinion on two related questions:

1　What range of management options should be followed – could this include non-intervention?
2　To what extent are the benefits likely to be economic and to whom?

The results of the research were collected and a SWOT analysis made to the second stakeholder meeting (see Table CS1.1).

All those attending the second stakeholder meeting were given a summary of the findings. There was general agreement for the vision that had been compiled from stakeholder inputs during the

TABLE CS1.1 **SWOT analysis of forest quality in one landscape in Wales**

Strengths	Weaknesses	Opportunities	Threats
Relatively high woodland cover	Domination by exotic species and plantations	New vision within the Welsh Assembly	Over-optimism about economic returns from forest management
High aesthetic and historical value	Fragility of much ancient woodland	Possible designation as biosphere reserve	
Remnant ancient forest	Lack of regeneration	Dyfi forest currently being redesigned by Forest Enterprise	Individual needs and aims over-riding community aims
Healthy trees	No old-growth forest		
Protected areas	Historic mismanagement		
Interest by government and government agencies	Low proportion owned locally	Development of new products and markets	Possible decline in Welsh tourist industry
Interest by landowners			
Good infrastructure (roads, sawmills etc.)	Welsh timber relatively unprofitable	More tourism: both specifically and peripherally related to woodlands	Tension between wider aims of government forest management and need to generate finance
Forest provides employment	Lack of biological knowledge		
Diversity of uses	Limited local markets	Opportunities for restoration	
Opportunities for public participation in management	Loss of woodland craft knowledge	Grants for a range of management options	Possible decline of Welsh farming
Footpath and cycle network	Poor farm economics	Community woodlands	
High tourism value	Different authorities in the catchment	Reassessment of land uses due to problems faced by upland farms	
Open to cooperation			
Some local markets			
Availability of expertise			

research: strong support for more broadleaves, a multipurpose forest estate and more benefits for local communities. A series of specific proposals were discussed:

- *A role in helping coordinate initiatives relating to a possible biosphere reserve:* the government is currently considering redesignating the catchment as a UNESCO biosphere reserve. If this occurs, it will probably include the full catchment, rather than being confined to the two National Nature Reserves as was the case in the past (this would no longer be acceptable as a Man and the Biosphere (MAB) reserve, which must include sustainable use areas as well). The Countryside Council for Wales has indicated that it would need a partner in management of any biosphere reserve and the DEVP has been suggested as a suitable partner. There was general support from the meeting both for the concept of a biosphere reserve in the catchment and for a role for the Partnership in its facilitation and management.

Your vision for Dyfi woodlands?

Come to a meeting to:

Hear an assessment of "Forest Quality" in the Dyfi
Valley

Create a community vision, and

Contribute to the Dyfi Eco Valley Partnership plans

Wednesday 28 June, 7.30 pm

Owain Glyndŵr Institute, Machynlleth

Simultaneous translation from Welsh to English *For further
information:* Andy Rowland 01654 705018

Coetiroedd y Fro - eich gweledigaeth?

Dewch i gyfarfod:

i glywed asesiad cychwynol ar "Ansawdd Coetiroedd"
Dyffryn Dyfi

i greu gweledigaeth cymunedol, ac

i gyfranu i gynlluniau Partneriaeth Eco Dyffryn Dyfi

Nos Fercher 28 Mehefin,7.30 y.h.

Canolfan Owain Glyndŵr, Machynlleth

Bydd cyfleusterau cyfieithu ar y pryd ar gael
Am ragor o wybodaeth: Andy Rowland 01654 705018

Note: Bilingual publicity material for the stakeholder meetings, which were also advertised through articles in the local newspapers.

FIGURE CS1.2 **Publicity material for the stakeholder meetings**

- *A community education role in terms of grants, commercial options, biodiversity, tourism and so on:* an immediate role was agreed for the Partnership in pulling together information and perhaps training courses for local people relating to existing opportunities, including grants, for woodland management, woodland conservation and new woodland planting.
- *Specific involvement in community-managed woodland in the catchment:* the particular role of managing a model community-managed woodland was discussed. While seen as a desirable option no firm conclusions were reached as to its viability.

- *Developing wood-fuel options*: several woodland owners or managers in the catchment are already managing their woodland to provide solid fuel, at present as wood. Some woods have been certified under the Forest Stewardship Council (FSC) scheme through the NGO Coed Cymru (although wood fuel cannot be labelled as certified by the Council at present). It was suggested that DEVP could coordinate greater use of fuelwood and perhaps the development of charcoal making or marketing within the catchment.
- *Coordination of a mapping and planning exercise leading to a community approach to forest management*: a more ambitious scheme is to continue the forest quality assessment into a more detailed mapping phase. This would be to identify and if possible negotiate uses of woodland between different owners and stakeholders in the DEVP area to produce a landscape approach to forest management and a catchment-wide approach to conservation, management, sharing of marketing options and so on. DEVP and the consultants agreed to develop the proposals.

LESSONS LEARNED

The methodology worked well enough to collect information and get a useful result in the time and resources available. The use of a workshop was not in itself sufficient – some groups did not speak enough and language may have been a problem, even though we had a translator present. The fact that the second meeting took place during harvest time meant that the farming community was poorly represented and in this case smaller meetings with specific interest groups might have been more representative. Plans to designate a UNESCO biosphere reserve are proceeding quite slowly and the forest vision will become more relevant if and when this designation occurs, although a long delay might mean that the process should really be repeated.

Case Study 2
A Partly Participatory Assessment in Cameroon) drawn from a larger study by Elie Hakizumwami)

Lobéké National Park is a 184,000 hectare area of forest in southern Cameroon, which was designated as a wildlife reserve in 1974 and redesignated as a national park in 2001. It is close to the borders of both the Central African Republic and Congo Brazzaville. Much of the protected area remains as primary forest although there has been extensive logging all around and, in the past, within the boundaries of what is now the national park itself. The area has little tourism, although there is a tented encampment and two walking trails; a major problem from a tourism perspective is the area's remoteness, usually taking two days to reach, mainly driving on dirt roads of varying quality. The assessment looked at the national park in the context of the wider landscape.

In Cameroon, initial work centred on the need to assess forest quality on an ecoregional scale, as part of the World Wide Fund for Nature (WWF) Ecoregion-based Conservation planning process in the Congo Basin. As a result, the scale of assessment is considerably larger than attempted in Europe. Given the area covered and the need for speed, the methodology presented below is also more 'top down' than in smaller landscapes, where local people are involved in the process of choosing indicators. Further developments will therefore be needed about ways in which local stakeholders can become involved in the larger-scale assessment.

The following steps were involved in the assessment:

- initial discussions with government officials and WWF staff in Yaoundé and on the site;
- agreement with staff of IUCN (The World Conservation Union) and WWF about the correct indicators to use on the site;
- discussions with other key stakeholders in the Lobéké area;
- literature review (drawing on published and unpublished reports, maps, papers etc.);
- field visits to collect information from staff and local communities;
- analysis using a SWOT (strengths, weaknesses, opportunities and threats) assessment;
- assessment of authenticity;
- discussion of results.

From the information collected during the survey, it can be concluded that the quality of the Lobéké forest remains fairly high from both a conservation and social perspective because of its considerable value, particularly in terms of biodiversity richness, regional ecological role, and the goods and services the forest provides for the development of national and local economies. Its socio-cultural value is also highly acknowledged by local communities.

SUMMARY OF RESULTS

The Lobéké forest is of considerable importance for any conservation strategy and still supports high densities of fauna and flora. Despite increasing threats to forest resources, many species internationally recognized as endangered continue to thrive in Lobéké. It is one of the few parts of the Guinea-Congolian ecosytem in Cameroon that still includes a significant proportion of primary forest, with unlogged areas. Lobéké provides potential for ecoregional and transboundary management of forest resources. It is contiguous with protected areas both in Central African Republic (Dzanga-Sangha National Park) and in Congo-Brazzaville (Nouabalé-Ndoki Forest Reserve), and is connected to Boumba-Bek and Nki by corridors allowing ecological exchange. The forest is a resource and sacred site to local communities who rely heavily on it for food, medicines and building materials, as well as cultural and spiritual identity. Their environment and lifestyles, however, are threatened by the activities of outsiders attracted to the region by ongoing intensive commercial logging and the opening-up of the forest.

Fundamental problems affecting the integrity of the Lobéké forest are related to the over-exploitation of some natural resources and the lack of effective management systems for the protection and sustainable use of these resources. The factors that contribute to unsustainable development include:

- conflicting interests of different stakeholders;
- lack of regulation to deal with the use of forest resources outside protected areas;
- ineffective government services for the control of natural resources in the area, particularly timber exploitation and poaching;
- remoteness of the area posing natural obstacles for effective monitoring of the use of natural resources, and limiting the tourism potential;
- widely vulnerable frontiers allowing cross-border poaching;
- an influx of outsiders seeking jobs in the area;
- the enclave nature of the indigenous population;
- poverty and a fragile economic environment;
- insufficient collaboration between key natural resource users;
- increasingly unsustainable commercial bushmeat hunting, destructive parrot trapping and uncontrolled fishing;
- uncontrolled timber exploitation, which in some areas is practised in an unsustainable way;
- increasing human pressure as a result of rapid local population growth;
- agricultural encroachment.

Conflicts recorded regarding the use of forest resources in the Lobéké area include:

- conflicts between local populations and poachers regarding the share of bushmeat and excess hunting;
- conflicts between local population and the government regarding the delimitation of the forest and access rights to the protected area;
- conflicts between economic operators (safari and logging companies) and local populations regarding the sharing of revenues generated from the forest and provision of jobs;
- conflicts between the local population and logging companies about the exploitation of the Ayous, which hosts caterpillars appreciated as a delicacy by local communities (conflict is still low);
- ethnic conflicts related to the forest ownership;
- boundary conflicts between the local population and logging companies, and between logging companies and local Ministére des forêts et de la faune (MINEF) of Cameroon, related to respect of logging concession boundaries and to the breaking of contracts between government and logging companies;
- conflicts between poachers and safari companies.

Local community activities are not at present having a serious negative impact on the global integrity of the Lobéké forest. However, the situation is likely to change in the near future with the influx of outsiders coming into the area looking for any kind of activities to generate cash income. In order to sustain the quality of the Lobéké forest, preventive measures need to be set up. In this context, local institutions need to be effectively empowered to ensure efficient monitoring of the use of natural resources in the area and the mobilization of different stakeholders towards sustainable use of natural resources. Free movement of indigenous people and consideration of communities customary territories during the use of forest resources reflect local communities' de facto landscape vision.

LESSONS LEARNED BY USING THE METHODOLOGY

The indicators chosen were easy to use for collection of the information necessary for an overall evaluation of the quality of the forest against the main criteria. Although some indicators were not adapted to the current situation, these may be maintained for use in the future according to the environmental and socio-economic dynamics in the region, including carbon sequestration, climate change and grazing.

Challenges with the assessment are associated with conflicting interests by different resources users or stakeholders. Given the connectivity of Central African forests, cross-border distribution of closely related human communities with traditional and free transboundary use of natural resources, along with free movements of animals, the assessment unit may need to be extended to an ecoregional level instead of being restricted to the forest block level. This will provide an opportunity for further analysis of a wide range of external factors while saving time on similarities. However, it would also increase the challenges of data collection and stakeholder involvement.

Case Study 3
Developing a Teaching Kit in Central America (drawn from a larger study by Alberto Salas)

Forest status varies dramatically between countries in Central America, with some states already having lost the majority of their forests, such as El Salvador, while countries like Nicaragua still contain extensive forest areas. Net forest loss continues in most countries. In the last two decades, there have been considerable efforts to develop protected area networks and sustainable forest management, the most ambitious plans being for a Meso-American Biological Corridor running through seven countries and including both protected areas and corridors of sustainably managed forests and other lands.

In this case, the forest quality assessment aimed to help monitoring of progress on the corridor and other conservation projects and, as in the case of Africa, was also applied ona large scale. There has until recently been little tradition of environmental monitoring in Central America. At the beginning of the assessment project, the regional forest criteria and indicator process – the Lepaterique Process – had stalled from lack of funding. The project supported development of a landscape-scale forest quality assessment method, drawing on experience from the Lepaterique process and the forest quality methodology. Following two regional workshops the forest quality methodology was finalized and a theoretical document and training manual were produced. PowerPoint presentations of the methodologies were completed. Training in forest quality assessment was carried out in five countries and around 200 people were trained (Herrera and Salas, 1999b; 1999c; 1999d).

TABLE CS3.1 **Matrix of training in Central America**

Country	Institutions involved
Costa Rica	Government agencies
El Salvador	Government agencies and NGOs
Guatemala	Government agencies and NGOs
Honduras	Government agencies and NGOs
Nicaragua	Government agencies and NGOs

Following training, there are continuing attempts to introduce the methodologies:

- In Costa Rica the government is planning to use the forest quality methodology within forested watersheds currently conserved for their role in provision of drinking water.

- In El Salvador, while there is interest in the methodologies, there is little culture of monitoring and evaluation in the country and one initial conclusion of the review carried out after training is that both the forestry and protected area departments are too weak to handle monitoring.
- In Guatemala a first landscape application of the two methodologies has taken place in and around Laguna Lachua National Park, a forest and freshwater protected area towards the north of the country.
- In Nicaragua the German technical cooperation organization, GTZ, agreed to fund testing of the methodology in Siapaz peace park and its buffer zone and this assessment is now complete. GTZ is interested in applying the methodology more generally.

LESSONS LEARNED

The methodology has been developed, along with training manuals, and training has taken place in several regions. However, actual testing on the ground remains incomplete at the present moment and while considerable progress has been made towards integrating these approaches with governments, the eventual success of this will only be judged after several more years. The process of introducing and explaining landscape-scale concepts of environmental quality proved difficult in places where many people in the population had only arrived recently, and where natural forest was still often regarded as an obstacle to be removed in order to establish farmland.

Case Study 4
Investigating Data Availability in a Swiss Canton (drawn from a longer study by Christian Glenz)

Swiss forest law already prioritizes many elements of sustainable forest management, including a preference for continuous-cover forestry, high importance given to biodiversity conservation and a strong role for protective forest management, particularly to mitigate avalanche damage.

An evaluation of the rapid assessment method was carried out in the forest district of the commune of Solothurn (Switzerland), in collaboration with Ruedi Iseli, the District Forest Officer. The total surface of the forest is about 2100 hectares, of which 1200 hectares are in the Jura Mountains and 900 hectares on the Swiss plateau, including 1630 hectares of exploited forest, 260 hectares of unexploited forest and 210 hectares of reserves. The forest district is distributed over 17 different communes, each of which will have some rights to timber, particularly for firewood. The forest ranges from an altitude of 430 to 1380 metres. Each of the indicators was analysed in turn to see if locally or nationally available data were available.

The study differed from others in focusing on the forest management unit level, which is an area of forest units that are not contiguous and are interspersed with other land. This caused problems in both collection and interpretation of data and led us to conclude that assessments are only likely to be valid if they can consider an entire landscape rather than unconnected fragments excised from a landscape.

However, the main interest from the current perspective is that the case study provides an opportunity to assess whether or not data are likely to be available in published literature. As will be seen from a summary of the results given in Table CS4.1, a large amount of information is *not* available, even in a rich country like Switzerland with a high level of research, suggesting that some rapid assessment method, expert workshop or data collection process will be needed.

TABLE CS4.1 **Data collection in the Solothurn area of Switzerland**

Indicator	Details
Authenticity	
Authenticity of composition	In most forest services, quantitative data is available through the forest inventory, including in Solothurn. Eight different tree species are distinguished (*Picea, Abies, Pinus, Larix, Fagus, Quercus, Acer* and *Fraxinus*) with others grouped as residual tree species. Data have been systematically collected through sampling (231 sample areas) every ten years since 1960. Data during the sampling period up to 1990 include coniferous tree species (60%: 38% *Picea*, 16% *Abies alba*, 3% *Pinus*, 2% *Larix* and 1% others) and deciduous tree species (40 %: 30% *Fagus*, 4% *Acer*, 3% *Fraxinus*, 2% *Quercus* and 1% others)

TABLE CS4.1 **Data collection in the Solothurn area of Switzerland** *(continued)*

Indicator	Details
	With respect to other species of fauna and flora, in Switzerland a complete biodiversity survey is normally not available. In a rapid assessment, there is insufficient time to do even a rapid biodiversity assessment, which takes almost several months (Groombridge and Jenkins, 1996). The most complete information is available at a national level (which is mostly an extrapolation using GIS), including information about mammals (Hausser, 1995), birds (Schmid et al, 1998) and plants (Welten and Sutter, 1982). At a landscape level data is missing, or collected only for some few species (e.g. wild ungulates). At a landscape level, a systematic survey of keystone and indicator species is not usually carried out. Data about abundance of the different wild ungulate species, as well as the other game species, were estimated by the hunters and their hunting guards. There is no real control of the abundance of these species by a communal or cantonal office and there is no guarantee of how representative this data is for Solothurn
	Presence of exotic species (animals or plants): No systematic surveys have been carried out. In the case of tree species, observations have been made with respect to the presence of some Douglas fir and similar species (less than 1% of the total (R. Iseli, 1999, personal communication))
	Ecosystems: Again no systematic surveys have been carried out. Data have to be obtained by remote sensing (photo-interpretation methods) or direct field observations. This indicator is more important when the methodology is used in a landscape with different land-use types rather than a forest district with a forest cover of 100%
Authenticity of structure	*Forest mosaic*: In Solothurn, trees have been classified by their development stages and trunk diameter, with information available since the 1960s
	Fragmentation and integrity: No official digitized map is available, but other cartographic material exists at a scale of 1:5000 and 1:25,000. To quantify fragmentation for example by a shape index (interaction of shape and size influence a number of ecological processes (Iorgulescu, 1997), digitized maps and information packages are needed. To use the fragmentation index as a way to represent forest quality is difficult in this case because not all the forest 'patches' are in reality delimited forest areas, because boundaries are only administrative and some parts remain contiguous with other forests in adjacency
Authenticity of function	*Viability of population*: Measured by presence of indicator species. No detailed data are available at landscape scale for either plant or animal species. Even if data were available for indicator species, it is doubtful if this would be sufficient to make conclusions about population viability, which needs modelling to factor in population dynamics (Barbault, 1995). Furthermore, the presence of an indicator species tells nothing about the status of the population
	Integrity of food webs: The measurement of this indicator would be time-consuming and not appropriate for this approach

TABLE CS4.1 **Data collection in the Solothurn area of Switzerland** *(continued)*

Indicator	Details
	Site characteristics: A map with geological data is available but other data has still to be checked
	Continuity of forest: Data are available, based on the forest inventory surveys periodically carried out since 1921. For more historical data other sources have to be checked
Authenticity of process	*Disturbance regimes*: According to the district forest officer, there have not been any natural or unnatural disturbance regimes (with the possible exception of localized fires). Deadwood is currently not quantified and in the past was even removed from the forest. In future, it is intended to retain deadwood in the forest because of requirements under certification by the Forest Stewardship Council (FSC)
Authenticity of area	*Continuity (age of forest)*: The only historical data currently available are forest inventory data collected since 1921. Other historical data should theoretically be available, for example from local archives, however tree ring analysis and pollen analysis have never been carried out for this area
	Integration: The study area is at a scale of a forest management unit and therefore adjacent forests have to be taken in consideration, otherwise there would be a misinterpretation of the real state
	Ratio of other land types: Data about land use are available from the Swiss federal statistic office at a resolution of 1 hectare. As only the forest district is considered and not the whole landscape, forest is the predominant land-use type. Adjacent to the forest, the main land-use types are pastures and arable lands
	Connectivity: There are protected areas, but their connectivity has not been verified. The configuration of this study area makes application of this indicator difficult as it needs to be evaluated at a larger scale, including the adjacent landscape
Authenticity of robustness and resilience	*Tree health*: No systematic survey has been made. The only data available is for the yearly increment and the quantity of insecticides applied (in spring for treatment of deadwood)
	Threats to tree health: Again, no systematic survey has been carried out. No information is available about levels of pollutants nor about any polluting activity in the forest and little information is available about invasive species. The only pest in these last years was the bark beetle. Bark-beetle traps have been installed
Development patterns	*Presence of human disturbance*: In the forest of Solothurn over 100 kilometres of roads exist, also used by trucks. A separate 'access map' (1:5000) is available
Environmental benefits	
Biodiversity	*Presence of representative species*: This has already been covered above (and repetition could distort quantitative assessments)

TABLE CS4.1 **Data collection in the Solothurn area of Switzerland** *(continued)*

Indicator	Details
	Presence of important biodiversity: Indication about Red List species status is given in the *Schweizer Brutvögelatlas* (Schmid et al, 1998) and for the mammals the list could be compared to the *Atlas of the Mammals*. For the other species (non-vertebrates) information is currently missing. There are no indications about endemic species (M. Baumann, 1999, personal communication). A special case is the presence of the wild cat (*Felis silvestris*)
Protected areas	*Presence of protected areas*: In the forest 200 hectares are protected as reserves and managed at the level of the canton of Solothurn. There are several other reserves delimited by the district forest officer. These reserves are variable in size and geographical location
	Effectiveness of protected areas: There is no management plan for the permanent forest reserves, although according to the district forest officer they are free of human intervention. The other forest reserves can in contrast partially be exploited. Information about the infrastructure is available from the access map (1:5000)
Watershed protection	*Official protection*: There are apparently no protected watersheds
	Effectiveness of protection: Data about soil quality have not been collected systematically overall in the canton. Only sporadic studies of forest soils are available
Climate issues	*Global climate change*: No data available
Social and economic benefits	
Wood products	*Amounts of timber and other wood products*: Based on the forest inventory, this data is available for every year
	Types of timber: As above
	Value: Based on the income statement, data are available for monetary value, however for non-monetary value, no indications exist
Non-timber forest products	*Type/Volume/Value/Areas for grazing*: No data exists
Employment	*People employed*: Very good data are available concerning the number of people employed and the periods of employment. From 1961 to 1998, the number of employed people declined from 41 to 12.5 due to increased mechanization and the lack of financial sources. In contrast, information about people indirectly employed through associated activities is not available
Value	*Monetary value of employment*: Data is available on the basis of the income statement
Subsistence activities	*Number of people using the forest*: No information is available

TABLE CS4.1 **Data collection in the Solothurn area of Switzerland** *(continued)*

Indicator	Details
Social values	*Proportion of people employed/Working conditions*: Good information is available, especially for the proportion of local people employed
Recreational value	*Number and type of visitors/Value of recreational value to local community/Infrastructure*: The only information relates to the number of specific tourist facilities
Homeland	*Number and type of people living in the forest/Land tenure issues/ Stability*: There are no settlements and no people living in the forest. As we consider only the forest district, no other types of land rights exist. The land belongs to the commune of Solothurn
Historical and cultural value	*Cultural artefacts and landscape*: There are some cantonal protected glacier relics and a ravine called the Verena Ravine. More data are available at the forest service of the canton of Solothurn and maybe in the book *Archeologie und Denkmalpflege im Kanton Solothurn* published in 1998
Aesthetic value	*Official recognition of aesthetic value/Other measures*: Has as yet never been considered
Educational value	*Importance and current use*: There are no real educational activities except for the 'learn trial'
Spiritual/religious value	*Existence of spiritual sites/Importance of sites*: The only known site is the Verena Ravine, which is an Anchorite site and is frequently visited
Local distinctiveness	*Areas of high value to local people*: No interviews have been made to identify local site values
Management and land use	*Type of management/Opportunities for the stakeholders' participation*: The total surface of the forest district is certified (FSC). Information about the surfaces of the different protected areas and their handling is available. The forest management plan is under construction. The forest is the property of the commune of Solothurn and therefore no other persons are implied in the discussions about the forest management investigations *Ownership of forest and forest resources/Access and use rights/ Intellectual property rights*: The owner of the forest is the commune of Solothurn, which also maintains the rights to use timber. Concerning access rights, everybody has free access to the forest, because of 'Swiss law for free access'. As already cited, wood resources are exploited by the commune of Solothurn, therefore the wild ungulates and the other game animals are exploited by the hunter association. Other forest products can be used by everybody
Knowledge	*Knowledge*: There is a high understanding relating to forestry
Nature of incentives	*Nature of incentives for particular types of forest management*: All parties are considered to be of equal importance under Swiss law
Institutions	*Types and nature of institutions affecting the forest landscape*: Good data quality available

LESSONS LEARNED

Data for most of the indicators are therefore apparently not available, with the exception of forestry statistics. If this is true for a well-developed country such as Switzerland, with a history of research, strong government infrastructure and well-funded forestry departments, it is likely to be even truer for many developing countries. The results of this field test were a major component in the decision to draw up the framework for the rapid assessment system as a combination of literature research and interviews.

Case Study 5
A Monitoring and Evaluation System in Viet Nam

The Central Truong Son is an incredibly diverse landscape of moist evergreen forests, karst limestone forests, open grasslands, upland plateaus and wetlands. These ecosystems support a variety of wildlife species, some still abundant and others already extremely rare. The area is also home to thousands of upland and lowland human communities who cultivate the land, utilize forest products and depend on water resources (Baltzer et al, 2001). The Central Truong Son Initiative was established to address the urgent threats to biodiversity in the region. It is a joint project between the government, non-governmental organizations (NGOs) and donors and covers eight provinces in central Vietnam.

One important element in the Central Truong Son Action Plan is development of a monitoring and evaluation system to measure progress on the action plan and in terms of key outcomes. The forest quality process was used to developed a long-term monitoring and evaluation system for forest landscape restoration in eight provinces in the Central Truong Son, part of the Annamite range of mountains in Viet Nam (Dudley et al, 2003). The monitoring and evaluation system had a series of aims:

- monitor progress on the Central Truong Son Initiative Action Plan;
- measure trends in environmental and social factors;
- help in communicating the initiative's achievements;
- provide information to help with adaptive management;
- give early warning of potential problems;
- lead to greater understanding of what local people want from the landscape;
- supply data for long-term research.

A long-term monitoring system requires a very different approach from the development of a one-off assessment, although in both cases the agreement of indicators is itself part of a negotiation process about what the landscape should support.

Viet Nam has a comprehensive and regularly maintained information system on many issues relating to agriculture, population, economic status and some aspects of natural resources. Alternatively, few other institutions outside the government collect information on a national scale. Some serious gaps in information remain, particularly on the issues most specific to conservation. There is virtually no information on forest condition, approach to forest management, natural regeneration, protected area effectiveness, or on status of biodiversity including threatened mammals and birds; the simplest baseline data on biodiversity does not exist for most of the region and new species are still regularly discovered. In addition, existing surveys tend to be of measurable *facts* rather than of *opinions* of stakeholders, yet both are important to management.

Development of the Central Truong Son monitoring and evaluation system was a cooperative exercise. Over 60 meetings took place with stakeholders at national, provincial, district and commune level to identify a small number of core indicators, which form the backbone of the system. Indicators measure progress on four different issues, against the context of threats to the Central Truong Son and its biodiversity:

- forest condition and biodiversity;
- forest ecosystem services;
- livelihoods;
- capacity for good management of natural resources;
- threats.

In addition to the core indicators, a smaller number of *flagship indicators* were suggested to provide key ways of measuring progress over time and the annual analysis will also include additional information on key research and surveys undertaken during the previous year.

Many indicators come from existing government statistics, sometimes with extra analysis, and some additional indicators will be monitored by World Wide Fund for Nature (WWF) and other stakeholders. WWF will monitor stakeholder attitudes to conservation for instance and perhaps progress in sustainable forest management, while the World Bank is monitoring management effectiveness of protected areas. Indicators are classified according to the ease of collection; some imply a level of capacity building and resources and may not be possible immediately. The project has been in discussions with the Forest Sector Support Programme to see if the work could be applicable on a wider scale in Viet Nam.

Proposed core indicators include the following:

- area of natural forest;
- forest quality;
- area of plantations;
- timber products (legal and illegal);
- non-timber forest products;
- sustainable forest management;
- amount of certified forest;
- percentage of reforestation budget for natural regeneration;
- number of natural forest regeneration projects;
- area judged to need restoration;
- number of forest fires;
- extent of forest fires;
- number of wildlife restaurants;
- wildlife trade from key ports;
- population of target species in protected areas;
- area of target habitat in protected areas;
- protected areas (number and location);
- protected area effectiveness;
- number of protected area management boards;
- catchment protection;
- irrigation enhancement;
- life expectancy by income class;
- access to family planning;
- access to health centres;
- access to electricity;
- percentage of boys/girls in secondary school;
- percentage of settled families;
- local stakeholder opinions;
- government protected area (PA) staff attending training courses;
- number of arrests for illegal hunting by guards;
- number of arrests for wildlife trade by guards;
- number of communes with volunteer rangers;
- achievement of Central Truong Son Initiative;
- kilometres of road in the eight provinces;
- total human population;
- impacts of the Ho Chi Minh highway.

Indicators are categorized according to whether information is already available, available with a little extra work or only available with considerable capacity building. The initiative will coordinate data collection from these different sources, wherever possible by linking databases. Core indicators will be augmented by additional information culled from research reports and field surveys so that trends will be set within a richer picture drawn from the increasing knowledge of the region. Benchmarks are also suggested for each of the core indicators to provide a target to assess against; in some cases these require further discussion by the Central Truong Son Initiative.

Succinct annual *State of the Central Truong Son* reports will summarize data, trends and key issues and information emerging over the previous 12 months. A draft table of contents for the report is suggested:

- Cover: title, illustration and box with key finding from the year;
- Summary table of flagship indicators, all core indicators and other key information;
- Detailed analysis of the core indicators and supplementary information by section, including identification of management responses and further action needed;
- Key overall assessment of progress on outcomes;
- Analysis of progress on the Action Plan;
- Details of the Central Truong Son Initiative and monitoring system.

The aim is for a system that is not reliant on constant cash injections or large amounts of time and can continue for many decades.

Case Study 6
Assessing Authenticity in the Avon Gorge, UK

The Avon Gorge, running beside the city of Bristol in the UK, is a highly significant site for biodiversity in the country and has been famous for its plants since William Turner found honewort (*Trinia glauca*) there in 1562. There are over 500 plant species present including many national rarities and two globally endemic species of whitebeam (*Sorbus bristolienses* and *Sorbus wilmottiana*). The gorge is predominantly of carboniferous limestone. It is surrounded on one side by Clifton Downs, a mainly managed area of grassland used as recreational land by the city, and on the other by Leigh Woods, a protected area and the site of the current survey. Being directly next to the city the area is heavily used and the woods contain many walking and cycling paths. The woodland is almost entirely secondary, the area having in the past been grazed by sheep. Indeed, the level of tree and shrub encroachment and the virtual absence of grazing animals (there is a small population of deer) threatens the survival of some rare plants that rely on open grassland habitats. Along with encroachment, the existence of exotic plants, including herbaceous garden escapees and trees such as evergreen oak (*Quercus ilex*) create further problems (Green at al, 2000).

The woods would probably register as medium authenticity in the simple typology given in this book. They are predominantly of natural species, albeit with some aliens, and are largely unmanaged, although the protected area managers carry out some sanitary felling to prevent cliff erosion, and some coppicing. The datasheet for recording forest authenticity was tested out in Leigh Woods and the original questions modified considerably in light of this experience. In particular, additional optional answers were added to several questions and an overview question added to each of the main criteria (these have been incorporated in the version printed in this manual). The results for the Avon Gorge are given in table CS6.1. The answer in each case is picked out in **bold** in the table.

TABLE CS6.1 Data card for assessment of forest authenticity in the Avon Gorge

Data card for stand-level assessment of forest authenticity:

Indicator	Elements				
Composition	How natural is composition of tree species?	Fully natural	**Mainly natural**	Many exotics	Exotic
	How natural is composition of other species?	Fully natural	**Mainly natural**	Many exotics	Exotic
	Are alien species present?	**Significant or invasive aliens**	Some non-invasive aliens	No significant aliens	
	Overall authenticity of composition	Fully natural	**Mainly natural**	Significant exotics	Almost all exotic

Notes on composition: mainly natural because of large number of native species, including important tree species but significant changes in places due to exotics

Pattern	What is the tree age distribution?	Mixed including old	**Mixed middle age**	Mixed mainly young	Trees recently lost	Mono-culture
	Is the forest canopy natural?	Fully natural		**Mainly natural**		Mainly unnatural
	Size of the forest in hectares	620 hectares				
	Overall authenticity of pattern	Fully natural	**Mainly natural**	Significant alteration	Monoculture	

Notes on pattern: management continues to remove some old trees even in protected areas

Functioning	Are viable populations of resident plant and animal species present?	All viable	Most viable	**Many not viable**	Most not viable
	What are the soil characteristics?	Stable		**Limited erosion**	Serious erosion
	What are hydrological characteristics?	**Healthy**		Limited problems	Serious problems
	How much deadwood is present?	Natural amounts		**Limited amounts**	Virtually none
	Overall authenticity of functioning	Fully functioning	Mainly functioning	**Significant loss of functioning**	Not functioning in a natural way

Notes on functioning: many rare species are at too low populations to survive long term; small populations of grazing animals mean forest cannot function in a wholly natural manner

Process	Does a natural disturbance regime exist?	Wholly natural	Partly natural	**Mainly unnatural**	Wholly unnatural
	Does an unnatural disturbance regime exist?	List unnatural disturbance factors: management interventions including felling and removals			
	Overall authenticity of process	Wholly natural	Partly natural	**Mainly unnatural**	Wholly unnatural

TABLE CS6.1 Data card for assessment of forest authenticity in the Avon Gorge *(continued)*

Notes on process:

Continuity	Age (approximate length of continuous forest cover): in most of the area less than 100 years, some areas of older woodland				
	Are the forest edges natural or artificial?	Wholly natural	Mainly natural	**Quite unnatural**	Wholly unnatural
	Is the forest connected to other similar habitat?	Wholly connected	Still well connected	**Some limited connections**	Isolated
	Overall authenticity of continuity	Wholly natural	Partly natural	**Mainly unnatural**	Wholly unnatural

Notes on continuity: some fragments of ancient forest exist and are of high conservation value

Resilience	What is average tree health?	Good	**Average**		Poor
	What is the health of other environmentally sensitive species?	Good	**Average**		Poor
	Are there important introduced pests, diseases and invasive species?	List those that affect ecosystem health: Japanese knotweed (*Fallopia japonica*) is invasive in places			
	What are air pollution levels?	High	**Medium**		Low
	Overall authenticity of resilience	Wholly natural	**Partly natural**	Mainly unnatural	Wholly unnatural

Notes on resilience: at one time Leigh Woods was recorded as having high levels of damage that could be related to air pollution. Many epiphytic species of lichen and moss that would otherwise have been expected are no longer present. Air pollution levels are generally declining.

Case Study 7
Applying forest quality to management
– the landscape approach

This book is mainly about assessment. But assessments are of little use if they are not applied back into management; indeed one of the reasons that assessment systems fail or are abandoned is because they all too often become divorced from management priorities and take on a life of their own. We hope that forest quality assessments will always feed straight back into management decisions and not remain theoretical exercises.

The final case study is therefore slightly different. Forest quality assessment will generally be used as a means to effect change, particularly through the planning and implementation of management at a landscape scale (which will often actually mean many different management interventions by different actors throughout the landscape). Below we describe one way in which forest quality assessment has been integrated into wider landscape approaches to conservation.

During the course of developing a forest quality assessment system, its suitability for helping to negotiate conservation outcomes became apparent. From this a broader landscape approach has been developed that aims to balance the ecological, social and economic land uses necessary for sustainable development through negotiations among stakeholders. Practical manifestations of this are the integration of forest management and poverty alleviation attempted by IUCN (The World Conservation Union) (Fisher et al, 2005) and the integration of protection, management and of forests at a landscape scale, currently being implemented by WWF International (Aldrich et al, 2003).

The approach addresses questions that cannot be answered at a global level or site level, for instance how to:

- apply ecoregion conservation in forests (particularly in crowded landscapes);
- avoid the limitations of site- or target-driven approaches – for example how do we balance and find the synergy between forest certification and forest protected areas;
- reconcile top-down and bottom-up approaches to conservation planning and determine the right scale at which particular policy interventions should take place;
- address policy issues that cannot be generalized on a global scale – such as whether or not plantations help preserve biodiversity or provide social benefits;
- integrate different components of forest conservation, such as protection, management and forest landscape restoration.

A key assumption of the landscape approach is that land-use specialization impacts cannot be determined with any accuracy at site level but only within the landscape. Three principles deal with scale, specialization and trade-offs:

1 The larger the area, the greater the range of goods, services and natural processes that should be represented;
2 The larger the scale the less acceptable is a net loss of goods, services, natural processes and future options;

3 Specialization at a smaller scale should not impair the delivery of essential goods, services and natural processes at higher scales.

The landscape approach provides an assessment and negotiating framework that helps to both plan and then implement a collection of landscape-scale interventions to develop a forested landscape that fulfils multiple goals. It aims to help achieve conservation goals while at the same time balancing different needs within the landscape.

It can be used in several ways, for example:

- implementing regional conservation strategies;
- making policy decisions about interventions (such as the establishment of plantations or protected areas) that cannot be generalized at a global or regional level;
- helping position site-based conservation initiatives (for example protected areas) within the framework of surrounding land uses, social needs and political/economic aspirations.

The approach is based around a model outlined below, with the stages printed in bold relating most directly to the forest quality assessment methodology:

Agreement on project target

Stakeholder analysis to identify the range of expectations from the landscape

Agreement on the size and borders of the landscape under discussion

Assessment of current performance with respect to meeting agreed functions

Assessment of potential performance

Negotiation of a mosaic of land uses

Implementation of the agreed plans

Monitoring and feedback to encourage adaptive management

FIGURE CS7.1 **A framework for the landscape approach**

The landscape approach is itself a response to the CBD's call for *ecosystem approaches* to be developed. The landscape approach is now being applied in several of WWF's (World Wide Fund for Nature) focal forest ecoregions, including in Sichuan in China and in the Central Truong Son region of Viet Nam.

PART 4

Appendices: Broader Issues and Sources of Information

The appendices look at how the forest quality assessment fits with some other priorities of WWF and IUCN and provide some general reference material.

Appendix 1
A Forest Quality Scorecard

In the book we suggest that in some circumstances a scorecard approach could be used to summarize initial impressions about forest quality in a landscape. Such an approach would allow someone (or more than one person) with knowledge of the area to provide a very rapid overview of key elements of forest quality. The limitations of this are clear: accuracy and detail would almost certainly be compromised and a scorecard is certainly not enough to provide information for detailed planning or negotiation of interventions. It might however provide sufficient detail to allow rapid comparison of different conservation landscapes to refine further an ecoregional assessment or, with further refinement, a way of tracking progress on landscape-level forest conservation projects. While scorecards will probably have to be developed on a case-by-case basis, an initial draft is presented below.

A Forest Quality Scorecard		
Indicators of authenticity		
How much natural forest exists in the landscape? How is it distributed? Is authenticity stable, increasing or decreasing?		
Distribution of natural forest: divide forests within the landscape between the five categories of authenticity below (for further details of the typology, see Chapter 5)		
Authenticity	Description	% in landscape
Very low	Few natural species or ecological functions. Narrow range of seral stages and simplified structure	
Low	Highly modified forest. Limited range of possible species, often exotics present. Narrow range of seral stages. Limited size of trees and little continuity over time	
Medium	Reasonably natural forest but with some components highly modified; variable size and continuity over time	
High	Forests approaching the natural state but with some key elements reduced or missing – e.g. oldest forest or some species	
Very high	Near natural forests with little human disturbance or management; all seral stages present or potentially present	
Components of authenticity: answer the following questions about composition of forests		
Composition	Tick the boxes below that most closely resemble the situation within the landscape	
Natural species	Species tending to be stable or increasing	Species tending to be decreasing
All or almost all the expected species of plants or animals present		
Most of the expected species of plants or animals present		

FIGURE A1.1 **A draft scorecard for forest quality assessment**

Many of the expected or original species have disappeared from the landscape		
Invasive species	Stable or decreasing	Increasing
Invasive species have a major impact on the ecology of the landscape		
Invasive species have a moderate impact on the ecology of the landscape		
Invasive species are absent or have little impact on the ecology of the landscape		

Pattern	Tick the box below that most closely resembles the situation in the landscape		
Status	Natural pattern increasing	Natural pattern stable	Natural pattern decreasing
Forest and woodland in most or all of the landscape have a natural structure and canopy			
Forest and woodland in about half the landscape have a natural structure and canopy			
Forest and woodland in a small area of the landscape have a natural structure and canopy			
Forest and woodland in the landscape does not have a natural structure and canopy			

Function	Divide forests in the landscape into the categories below
Status	% in landscape
Most or all expected ecosystem functions and microhabitats occur	
Many expected ecosystem functions and microhabitats occur	
Few expected ecosystem functions and microhabitats occur	

Process	Divide forests in the landscape into the categories below
Status	% in landscape
Forest regenerates almost entirely through natural processes of disturbance patterns and seeding	
Forest regenerates fairly naturally but some large-scale disturbance patterns are absent	
Forest regenerates through planting or other forms of regeneration	

Resilience	Divide forests in the landscape into the categories below
Status	% in landscape
Forest almost or completely free of anthropogenic impacts on ecosystem health (air pollution, invasive pests and diseases, climate change)	

FIGURE A1.1 **A draft scorecard for forest quality assessment** *(continued)*

Forest susceptible to anthropogenic impacts on ecosystem health but still fairly free of symptoms		
Forest undergoing detrimental effects of anthropogenic impacts on ecosystem health		
Continuity	Divide forests in the landscape into the categories below	
Status		% in landscape
Forest has been present on the site for as long as records exist		
Forest has been present on the site for at least 200 years		
Forest has been present on the site for less than 200 years but more than 50 years		
Forest established less than 50 years ago		
Development patterns	Divide forests in the landscape into the categories below	
Status		% in landscape
Forest authenticity being compromised or reduced as a result of development patterns		
Forest authenticity unaffected by development patterns		
Forest authenticity being increased as a result of deliberate management actions		
Indicators of environmental benefits		
What environmental benefits do forests provide? Are they increasing or decreasing?		
Biodiversity	Divide forests in the landscape into the categories below	
Status		% in landscape
Forest in protected areas in IUCN categories I–IV		
Forest in protected areas in IUCN categories V–VI		
Forest with no official biodiversity protection status		
Soil and watershed protection	Divide forests in the landscape into the categories below	
Status		% in landscape
Forest set aside for watershed protection		
Forest with a strong watershed protection function but with no official status – currently stable		
Forest with a strong watershed protection function but with no official status – currently threatened or decreasing		

FIGURE A1.1 **A draft scorecard for forest quality assessment** *(continued)*

Impacts on other ecosystems	Tick the relevant box or boxes below	
Mangroves present in the landscape		
Riverine forests or river island forests present in the landscape		
Savannah edge or tundra edge forests present in the landscape		

Climate stabilization	Tick the box if relevant	
Carbon sequestration projects funded under the auspices of the Framework Convention on Climate Change present in the landscape		

Indicators of other social and economic benefits

What are the social and economic benefits derived from the forested landscape?

Wood products	Tick the relevant box below			
Status		Increasing	Stable	Decreasing
Commercial wood and/or fibre products are a major product from the landscape				
Commercial wood and/or fibre products are a minor product from the landscape				
Commercial wood and/or fibre products are not produced from the landscape				

Non-wood products	Tick the relevant box below			
Status		Increasing	Stable	Decreasing
Non-timber forest products for sale are a major product from the landscape				
Non-timber forest products for sale are a major product from the landscape				
Non-timber forest products for subsistence are of major importance in the landscape				
Non-timber forest products for subsistence are of minor importance in the landscape				

Employment and subsistence	Tick the relevant box below			
Status		Increasing	Stable	Decreasing
Full-time employment in the forest products industry is a major employer in the area				
Full-time employment in the forest products industry is a minor employer in the area				

FIGURE A1.1 **A draft scorecard for forest quality assessment** *(continued)*

	Increasing	Stable	Decreasing
There is no full-time employment in the forest products industry			
Part-time employment in the forest products industry is a major employer in the area			
Part-time employment in the forest products industry is a minor employer in the area			
There is no part-time employment in the forest products industry			

Recreation	Tick the relevant box below		
Status	Increasing	Stable	Decreasing
Forest based recreation is a major employer in the area			
Forest based recreation is a minor employer in the area			
There is no forest based recreation in the area			

Homeland	Tick the relevant box below		
Status	Increasing	Stable	Decreasing
Many people live in or around the forest and rely on it for subsistence			
A few people live in or around the forest and rely on it for subsistence			
No people live in or around the forest and rely on it for subsistence			

Historical values	Tick the relevant box or boxes below	
Status	Stable	Threatened
There are important historical sites within the forest landscape		
There are important historical forest management systems within the forest landscape		

Cultural and artistic values	Tick the relevant box below	
The landscape is recognized as having important cultural and/or artistic values which are stable		
The landscape is recognized as having important cultural and/or artistic values which are threatened		

Spiritual values	Sacred trees, sacred groves, burial sites, use for spiritual fulfilment

FIGURE A1.1 **A draft scorecard for forest quality assessment** *(continued)*

There are important spiritual sites within the forest landscape and these are protected	
There are important spiritual sites within the forest landscape and these are not protected but stable	
There are important spiritual sites within the forest landscape and these are not protected and under threat	

Management and land use	Tick the relevant box below	
Management is on the whole increasing overall forest quality		
Management is on the whole maintaining overall forest quality in a stable state		
Management is on the whole decreasing overall forest quality		

Rights and legal issues	Tick the relevant box below	
Access rights and land tenure are secure and understood for almost all or all of the forest landscape		
Access rights and land tenure are secure and understood for some but by no means all of the forest landscape		
Access rights and land tenure remain uncertain or in dispute for almost all or all of the forest landscape		

Knowledge	Tick the relevant box or boxes below	
The area is well known and understood by traditional and local peoples		
The area is well known and understood, at least in part, by the scientific community		
The area is well known and understood, at least in part, by the forest industry		

Nature of incentives	List any relevant incentives (good or bad) for forest management in the space below

Local distinctiveness	List any specifically local values within the forested landscape in the space below

FIGURE A1.1 **A draft scorecard for forest quality assessment** *(continued)*

Appendix 2
Links to forest landscape restoration

Aspects of forest quality assessment have been used to help draw up a monitoring and evaluation framework for forest landscape restoration. IUCN (The World Conservation Union) and WWF (World Wide Fund for Nature) define forest landscape restoration as: '*a planned process that aims to regain ecological integrity and enhance human well-being in deforested or degraded forest landscapes and beyond*', a definition drawn up at a workshop in Segovia, Spain in July 2000 (WWF and IUCN, 2000). It was agreed that choices about restoration should be made at a landscape scale and on a case-by-case basis in response to specific conditions – with overall landscape benefits being the goal against which success is measured.

Forest landscape restoration aims to address both socio-economic needs (such as ecotourism, sustainable timber production and livelihood security) and ecological needs (such as habitat, connectivity and soil protection). This inevitably involves trade-offs between different site-level functions, and key stakeholders should be involved in determining how to balance the trade-offs required for sustainable solutions (Mansourian et al, 2005).

AUTHENTICITY (NATURALNESS OR ECOLOGICAL INTEGRITY) OF FORESTS SHOULD INCREASE AT A LANDSCAPE SCALE

The underlying philosophy of forest landscape restoration is to favour natural regeneration over conventional tree planting, and also to favour management systems that involve minimum interference with the natural ecological cycle. An important underlying theme of the approach is that the more natural forest mosaic that results should have improved resilience to threats such as climate change and disturbances such as fires and storms. Nonetheless, within the landscape some sites may be dedicated to highly unnatural tree cover (wood fuel plantations, tree crops and so on) if these contribute to overall human well-being and do not become a dominant feature at a landscape scale.

ENVIRONMENTAL BENEFITS SHOULD AT LEAST REMAIN STABLE AT A SITE SCALE AND SHOULD INCREASE AT A LANDSCAPE SCALE

Forest management that results in either on-site or off-site environmental damage – such as soil erosion, fertilizer run-off, pesticide spray drift or downstream hydrological effects – is incompatible with the wider aims of forest landscape restoration. Thus the principles for environmental benefits are more stringent than for either authenticity or social benefits, having both site and landscape components. There may, however, be occasions when the best that can be hoped for environmental benefits at site level is that these will remain stable, so that the principle at site level is for no further decline. At a landscape scale, on the other hand, restoration and ecological resilience should result in an increase in environmental benefits.

LIVELIHOODS SECURED AT A LANDSCAPE SCALE

As with authenticity, forest landscape restoration may not improve social well-being at every site. But the definition is clear that these should improve on a landscape scale. The involvement of key stakeholders in decision-making processes should help to ensure that issues relating to human well-being are fully addressed. As forest landscape restoration provides a vehicle to halt and reverse forest loss and degradation, a key element in the approach is to address the underlying causes that drive forest loss. Many of these are linked to human well-being and include issues outside traditional conservation concerns, such as gender, equity and land tenure. Actions that aim to reverse the underlying causes of forest degradation at a landscape scale are of necessity long term and require matching long-term commitment from the various partners.

Criteria and generic indicators for forest landscape restoration

The variable nature of forest landscape restoration means that, although we can identify some criteria and a few general indicators, most indicators will need to be chosen on a case-by-case basis, chosen to fit within the template during project planning. A draft set of ten criteria and some examples of indicators are given in Table A2.1.

TABLE A2.1 **Some criteria and indicators of forest landscape restoration**

Criteria	Examples of specific indicators
Indicators relating to authenticity	
Forest composition and pattern	• Amount/proportion of natural forest (i.e. forest made up of natural species and allowed to develop natural characteristics) • Proportion of forest containing several different successional stages (measured against natural forest type of the region)
Forest ecosystem function and process	• Distribution of rare or threatened forest-dependent species • Amount of a specific indicator associated with natural forest processes – e.g. dead timber
Forest fragmentation and extent	• Area of forest in the landscape compared with original forest extent (use FAO definition of forest) • Median size of forest stands
Indicators relating to environmental benefits	
Environmental services	• Water quality and quantity • Changes in stream sediment load*
Environmental resilience/resistance	
Indicators relating to secure livelihoods	
Increased livelihood opportunities	A proxy measure of food, shelter, clothing, education etc., e.g.: • number of jobs supported by forests in the landscape; • numbers of key NTFPs available on a sustainable basis

TABLE A2.1 **Some criteria and indicators of forest landscape restoration** *(continued)*

Criteria	Examples of specific indicators
Reduced human vulnerability	Need indicators relating to specific 'pressure points' within a landscape
Increased equity	Specific indicators will be needed relating to targets in a landscape, e.g.: • number of traditional livelihoods supported; • opportunities for participation in management decisions
Maintenance of cultural values	Specific indicators will be needed relating to targets in a landscape, e.g.: • protection/restoration for sacred sites in forests; • number of recreational visits to forests and woodland
Enabling political and institutional environment	• Enabling legislation • Funding • Positive government incentives

Note: *While carbon sequestration might seem to be an ideal indicator, any use of this would require careful handling to ensure that WWF's position on the Kyoto Protocol of the Framework Convention on Climate Change is not undermined.

Appendix 3
Links to High Conservation Value Forests

The forest quality assessment approach developed some guidance about how different elements of quality might be measured, but did not attempt to rank these or to identify the elements of the 'high quality' forest. To some extent this is inherent in the methodology because it is predicated on the principle that there are many different interpretations of 'quality' and one aspect of landscape-scale management should be the integration of these. However, there are clearly reasons for different interest groups to have the means to identify their own understanding of 'high quality'. At one time there were plans to develop the forest quality methodology to allow this, but in the meantime the concept of High Conservation Value Forests (HCVF) had been developed, first by the Forest Stewardship Council (FSC) and then in much greater detail by the consultancy ProForest, working with World Wide Fund for Nature (WWF) International (Jennings et al, 2003). Given that the HCVF concept now provides a viable way of identifying quality from a conservation perspective, there seems to be little point in repeating this exercise.

ProForest identifies six values that separately or together identify HCVFs:

- *HCV1* – forest areas containing globally, regionally or nationally significant concentrations of biodiversity values (for example endemism, endangered species, refugia). *For example, the presence of several globally threatened bird species within a Kenyan montane forest.*
- *HCV2* – forest areas containing globally, regionally or nationally significant large landscape level forests, contained within, or containing the management unit, where viable populations of most if not all naturally occurring species exist in natural patterns of distribution and abundance. *For example, a large tract of Mesoamerican lowland rainforest with healthy populations of jaguars, tapirs, harpy eagles and caiman as well as most smaller species.*
- *HCV3* – forest areas that are in or contain rare, threatened or endangered ecosystems. *For example, patches of a regionally rare type of freshwater swamp forest in an Australian coastal district.*
- *HCV4* – forest areas that provide basic services of nature in critical situations (for example watershed protection and erosion control). *For example, forest on steep slopes with avalanche risk above a town in the European Alps.*
- *HCV5* – forest areas fundamental to meeting basic needs of local communities (for example subsistence and health). *For example, key hunting or foraging areas for communities living at subsistence level in a Cambodian lowland forest mosaic.*
- *HCV6* – forest areas critical to local communities' traditional cultural identity (areas of cultural, ecological, economic or religious significance identified in cooperation with such local communities). *For example, sacred burial grounds within a forest management area in Canada.*

The HCVF approach provides a great deal of flexibility in terms of developing assessments at different scales. Some of the approaches outlined in this manual could be used in implementing HCVF assessments at a landscape scale.

Appendix 4
Forest quality and forest certification

While the forest quality assessment method outlined in this book aims at a landscape level, other actors have been looking at forest condition at the level of the stand. The Forest Stewardship Council (FSC) has in particular developed a series of *Principles and Criteria* for accrediting independent certifiers of forest products (http://www.fsc.org). A preliminary comparison is given below. It shows:

- *Authenticity* was traditionally not addressed in detail in the *FSC Principles and Criteria*, except through Principle 9. While this has been enormously strengthened through development of the HCVF principle, it still gives little guidance about the status of other forests within a certified area.
- *Environmental benefits* are addressed by FSC Principle 6, although issues related to local and global climate stabilization, and impacts on habitats worldwide, are not covered.
- *Social and economic benefits* are amongst the best covered, with two principles relating to land use and indigenous people, one relating to employment and one to wider forest benefits. However, issues such as recreation, aesthetic values, local distinctiveness, spiritual and religious significance and educational value are either not or only poorly addressed.

The *FSC Principles and Criteria* provide a minimum requirement and individual standards have considerably more stringent requirements. Some more detailed comparisons are outlined in Table A4.1.

TABLE A4.1 **Comparison between forest quality and the Principles and Criteria of the FSC**

Suggested criteria of forest quality	FSC: Strong correlation with forest quality criteria	FSC: Weak correlation with forest quality criteria
AUTHENTICITY		Principle 9: Maintenance of natural forest
ENVIRONMENTAL BENEFITS	Principle 6: Environmental impact	Principle 1: Compliance with FSC Principles
Biodiversity conservation	Principle 6: Environmental impact	
Soil and watershed protection	Principle 6: Environmental impact	
Impacts on other semi-natural habitats		Principle 9: Maintenance of natural forests

TABLE A4.1 **Comparison between forest quality and the Principles and Criteria of the FSC** *(continued)*

Suggested criteria of forest quality	FSC: Strong correlation with forest quality criteria	FSC: Weak correlation with forest quality criteria
Local climatic effects		
Global climate effects		
Effects on habitats worldwide		
SOCIAL/ECONOMIC BENEFITS		
Wood products	Principle 5: Benefits from the forest	
Non-timber forest products	Principle 5: Benefits from the forest	
Employment and subsistence	Principle 4: Community relations and workers' rights	
Recreation		
Homeland for people	Principle 2: Tenure and use rights and responsibilities	
	Principle 3: Indigenous peoples' rights	
Historical and cultural importance	Principle 9: Maintenance of natural forest	
Aesthetic values		
Spiritual and religious significance		
Local distinctiveness		
Educational value including scientific research		Principle 8: Monitoring and assessment

Appendix 5
A brief history of the forest quality concept

Criteria of forest quality were first proposed in a World Wide Fund for Nature (WWF) report on temperate and boreal forests, published in 1992 (Dudley, 1992). Originally, four criteria were suggested – authenticity, forest health, environmental benefits and other social and economic benefits – along with a series of indicators. The system was used primarily as a lobbying instrument in connection with WWF's campaign to raise the political profile of environmental problems in temperate and boreal forests. In the years since, the concept has been developed and used in several ways:

- *The concept of forest quality has been further refined and developed.* The overall definition of forest quality was explored at a workshop in the UK in 1993, in consultation with a wide range of UK-based stakeholders from government and non-governmental organizations (NGOs) (Dudley et al, 1993). Since then, specific elements have been developed in greater detail, with efforts focusing particularly on the concept of *authenticity* (Dudley, 1996).
- *The ideas have been integrated into the general forest policy work of WWF and IUCN* (The World Conservation Union). Forest quality has been accepted as an integral element in the IUCN/WWF global forest strategy, *Forests for Life* (Dudley, Gilmour and Jeanrenaud, 1996; Jackson et al, 2000).
- *The principles have been promoted to governments, international organizations and industry.* This has been achieved through written material, briefings, workshops and conference presentations (in Canada, the US, Switzerland, Finland, Slovakia, Spain, Latvia, Georgia, Cameroon, Gabon, Costa Rica, Indonesia). Overheads, a slide set, a trilingual booklet (WWF, IUCN, GTZ and EPFL, 1999) (in English, French and Spanish) and a trilingual PowerPoint presentation have been developed to help put the concepts across in an accessible form. The importance of forest quality has, since WWF's work began, been widely accepted within the international community.
- *Forest quality criteria have been used to influence other international forest policy processes* through expert meetings, preparation of papers and so on, including:
    - The *Convention on Biological Diversity (CBD)*, through writing a report commissioned by the CBD Secretariat (Convention on Biological Diversity, 1997a), by involvement in CBD expert working groups on a forest work programme and on key biodiversity indicators (Convention on Biological Diversity 1997b) and by writing background lobbying papers for WWF (WWF, 1996a) – the concept now features clearly in the CBD's work programme;
    - The *Commission on Sustainable Development, Intergovernmental Panel on Forests* and *Intergovernmental Forum on Forests* through preparations of lobbying documents and by advocacy in meetings (IUCN and WWF, 1996; WWF, 1996a, 1996b);
    - The *UN Forest Resource Assessment* through preparation of a background paper (Dudley and Elliott, 1996) and involvement in expert working groups in Finland and Switzerland. The new Temperate and Boreal Forest Resources Assessment specifically includes elements of forest quality based on this work;
    - The various regional *criteria and indicator processes* through lobbying before and during the

launch of the Pan European Process and involvement in the expert working group at the start of the Montreal Process (Dudley, 1995);

- The *Forest Stewardship Council* through extensive involvement by WWF in setting the FSC's *Principles and Criteria.*

- *The forest quality concepts have been further discussed and developed through cooperation with other research organizations.* Examples include the World Resources Institute (WRI), World Conservation Monitoring Centre, WWF's Living Planet Campaign, the Global Biodiversity Forum (Dudley, 1997b), the World Commission on Sustainable Development, the European Commission COST Action network, the Center for International Forestry Research and the European Forest Institute (Dudley and Jeanrenaud, 1997) and through presentations at international meetings (Dudley, 1997a; Dudley and Rae, 1998).

- *Early work has taken place to examine the options for wider criteria of environmental quality working at the same level* for example for organic agriculture, food, consumption (Stolton and Dudley, 1994) and so on.

- *The forest quality methodology has been developed as an assessment tool at a landscape scale, in cooperation with the École Polytechnique Fédérale de Lausanne, WWF and IUCN and tested in Europe, Central America and Africa.* Work included:
  - a series of workshops in Switzerland (WWF, IUCN and EPFL, 1998), Africa (WWF and IUCN, 1998) and Central America (Corrales, 1999; Herrera and Corrales, 1999; Herrera and Salas, 1999a);
  - a number of associated working papers to address the questions raised by the workshop (Dudley and Stolton, 1998);
  - field testing in these regions with subsequent reports (Dudley et al, 1999);
  - preparation of a field manual.

- *Aspects of the forest quality work are being carried on in WWF's Forests for Life target-driven programme specifically relating to:*
  - the Landscape Approach (see Case Study 7);
  - monitoring forest landscape restoration (see Appendix 2).

Appendix 6
Glossary

Many of the terms used in this report have been adapted particularly for the particular uses employed here. This short glossary includes definitions of some key terms.

Authenticity: used here as a measure of ecosystem function and more loosely of 'naturalness'. An authentic forest is defined as *a forest in which all expected ecosystem functions can continue to operate indefinitely.*

Biodiversity: biological diversity measured in terms of ecosystems, species and genetic variation within species.

Bioregion: a geographically related assemblage of ecoregions that shares a similar biogeographical history and thus has strong affinities at higher taxonomic levels (for example genera, families).

Criteria and indicators of forest quality: a collection of elements that together helps to assess forest quality.

Criterion: a major category of conditions or processes – quantitative or qualitative – that together helps define forest quality. A criterion is characterized by a set of related indicators.

Ecoregion: a large area of land or water that contains a geographically distinct assemblage of natural communities that first, share a large majority of their species and ecological dynamics; second, share similar environmental conditions; and third, interact ecologically in ways that are critical for their long-term persistence.

Environmental benefits: used here to denote benefits from forests in terms of environmental services such as watershed, soil and climate protection, biodiversity and the role of the forest in helping to maintain other ecosystems.

Forest: in this methodology the term 'forest' is used very generally and includes the habitat often referred to as 'other wooded land'.

Forest quality: the significance and value of all the ecological, social and economic components of the landscape.

Incentives: economic or social factors that help facilitate or determine particular courses of action – in this case particular forms of forest management.

Indicator: a measure – quantitative or qualitative – that provides useful information about a criterion.

Landscape: many definitions exist. In this context, we use: *a contiguous area, intermediate in size between an 'ecoregion' and a 'site', with a specific set of ecological, cultural and/or socio-economic* *characteristics distinct from its neighbours.* In practical terms, the 'landscape' here is the area designated for inclusion in a forest quality assessment.

Local distinctiveness: used here to describe the particular value of a forest to a local community, *beyond* easily measurable values such as timber extracted or recreational use. Local distinctiveness is rooted in personal or locally collective behaviour and is thus inherently difficult to measure.

Non-timber forest products: all other products from the forests, such as food, medicines, resin, cork and so on – some non-timber forest products also come from trees.

Non-wood goods and services: a broader term than non-timber forest products also including services such as environmental benefits and broader values such as recreational use and aesthetic importance.

Other social and economic benefits: used here to denote benefits directly accruing to humans *in addition to* environmental benefits. The word 'other' is used in the title to reflect the economic importance of many environmental benefits.

Rapid quality assessment: used here to describe an assessment system that can be undertaken quickly and cheaply; accuracy of results will be proportionately less than in the case of more detailed assessments.

Resilience: used here as a measure of positive ecosystem health, encompassing both health itself and ability to withstand environmental stress and change.

Spiritual values: the values associated with religious and spiritual values, including hard-to-measure personal spiritual values.

Some of the definitions given above (ecoregion, bioregion) are drawn from information supplied to the *WWF Ecoregion-based Conservation Workshop*, Washington DC, 12–15 January 1997.

References

Agnoletti, M. and Anderson, S. (eds) (2000) *Forest History: International Studies on Socioeconomic and Forest Ecosystem Change*, CABI Publishing, Wallingford, UK

Aldrich, M., Bubb, P., Hostettler, S. and van de Wiel, H. (2000) *Tropical Montane Cloud Forests*, Arborvitae special issue, WWF and IUCN, Gland, Switzerland

Aldrich, M., Belokurov, A., Bowling, J., Dudley, N., Elliott, C., Higgins-Zogib, L., Hurd, J., Lacerda, L., Mansourian, S., McShane, T., Pollard, D., Sayer, J. and Schuyt, K. (2003) *Integrating Forest Protection, Management and Restoration at a Landscape Scale*, WWF International, Gland, Switzerland

Anderson, J. (2005) *Access and Control of Indigenous Knowledge in Libraries and Archives, Ownership and Future Use*, American Library Association and The McArthus Foundation, New York, Columbia University

Anderson, J. E. (1991) 'A conceptual framework for evaluating and quantifying naturalness', *Conservation Biology*, vol 5, pp347–352

Angelstam, P. (1992) 'Conservation of communities: The role of edges, surroundings and mosaic structure in man-dominated landscapes', in Hansson, L. (ed.) *Ecological Principles of Nature Conservation*, Elsevier, London, pp9–70

Angelstam, P. and Mikusinki, G. (1994) 'Woodpecker assemblages in natural and managed boreal and hemi-boreal forest: A review', *Annales Zoological Fennica*, vol 31, pp157–172

Angermeier, P. L. (1994) 'Does biodiversity include artificial diversity?' *Conservation Biology*, vol 8, pp600–602

Angermeier, P. L. and Karr, J. R. (1994) 'Biological integrity versus biological diversity as policy directives', *Bioscience*, vol 44, pp690–697

Anttila, E. (1996) *National Experiences on Criteria and Indicators, Intergovernmental Seminar on Criteria and Indicators for Sustainable Forest Management, 19–22 August, Helsinki Finland*, Ministry of Agriculture and Forestry, Helsinki

Attiwill, P. M. (1994) 'The disturbance of forest ecosystems: The ecological basis for conservative management', *Forest Ecology and Management*, vol 63, pp247–300

August, P. V. (1983) 'The role of habitat complexity and heterogeneity in structuring tropical mammal communities', *Ecology*, vol 64, pp1495–1507

August, P. V., Pickett, S. T. A. and Thompson, J. N. (1978) 'Patch dynamics and the design of nature reserves', *Biological Conservation*, vol 13, pp27–37

Bader, P., Jansson, S. and Johnson, B. G. (1995) 'Wood inhabiting fungi and substratum decline in selectively logged boreal spruce forests', *Conservation Biology*, vol 72, pp355–362

Baker, W. L. (1992) 'Effects of settlement and fire suppression on landscape structure', *Ecology*, vol 73, pp1879–1887

Baltzer, M. C., Nguyen Thi Dao and Shore, R. G. (2001) *Towards a Vision for Biodiversity Conservation in the Forests of the Lower Mekong River Complex*, WWF and partners, Hanoi

Barbault, R. (1995) *Ecologie des peuplements: Structure et dynamique de la biodiversité*, Masson, Paris

Bass, S. and Hearne, R. R. (1997) *Private Sector Forestry: A Review of Instruments for Ensuring Sustainability*, International Institute for Environment and Development, London

Bass, S. and Mayers, J. (2001) *The Forest Governance Pyramid*, World Bank and WWF, Washington DC and Gland, Switzerland

Bennett, A. F. (1990) 'Habitat corridors and the conservation of small mammals in a fragmented forest environment', *Landscape Ecology*, vol 4, pp109–112

Bertello, P. (1998) 'Assessing ecosystem health in governed landscapes: A framework for developing core indicators', *Ecosystem Health*, vol 4, no 1, pp33–51

Bilby, R. E. and Bisson, P. A. (1998) 'Functioning and distribution of large woody debris', in Naiman, R. J. and Bilby, R. E. (eds) *River Ecology and Management*, Springer, New York, pp324–346

Binkley, D. and Brown, T. C. (1993) *Management Impacts of Water Quality of Forests and Rangelands*, USDA Forest Service, General Technical Report RM-239, USDA Forest Service, Fort Collins, Colorado

Bobiec, A. (2002) 'Living stands and dead wood in the Bialowieza forest: Suggestions for restoration management', *Forest Ecology and Management*, vol 165, nos 1–3, pp125–140

Borges, J. L. (1962) *Labyrinths*, New Directions, New York

Boorman, F. H. and Likens, G. E. (1979) *Pattern and Process in a Forested Ecosystem*, Springer-Verlag, New York

Borrini-Feyerabend, G. (1996) *Collaborative Management of Protected Areas: Tailoring the Approach to the Context*, IUCN, Gland, Switzerland

Borrini-Feyerabend, G. with Buchan, D. (eds) (1997) *Beyond Fences: Seeking Social Sustainability in Conservation*, vols 1 and 2, IUCN, Gland, Switzerland, www.iucn.org/themes/spg/Files/beyond_fences/beyond_fences.html

Bourassa, S. C. (1991) *The Aesthetics of Landscape*, Belhaven Press, London

Bowden, C. and Hoblyn, R. (1990) 'The increasing importance of restocked conifer plantations for woodlarks in Britain: Implications and consequences', *RSPB Conservation Review*, vol 4, pp25–31

Boyle, T. J. and Boontawee, B. (eds) (1995) *Measuring and Monitoring Biodiversity in Tropical and Temperate Forests: Proceedings of an IUFRO Symposium held at Chiang Mai, Thailand, 27 August – 2 September 1994*, Center for International Forestry Research, Bogor, Indonesia

Broekhoven, G. (1996) *Non-timber Forest Products: Ecological and Economic Aspects of Exploitation in Colombia, Ecuador and Bolivia*, IUCN, Gland, Switzerland

Brown, P. (1998) *Climate, Biodiversity and Forests: Issues and Opportunities Emerging from the Kyoto Protocol*, World Resources Institute in collaboration with IUCN, Washington DC

Brown, S. (2002) 'Measuring, monitoring and verification of carbon benefits for forest-based projects', in Swingland, I. R. (ed.) *Capturing Carbon and Conserving Biodiversity: The Market Approach*, The Royal Society and Earthscan, London

Bruijnzeel, L. A. (1990) *Hydrology of Moist Tropical Forests and Effects of Conversion: A State of Knowledge Review*, UNESCO International Hydrological Humid Tropics Programme, Paris

Bruijnzeel, L. A. (2001) 'Hydrology of tropical montane cloud forests: A Reassessment', *Land Use and Water Resources Research*, vol 1, pp1.1–1.18

Bryant, D., Nielsen, D. and Tangley, L. (1997) *The Last Frontier Forests: Ecosystems and Economies on the Edge*, World Resources Institute, Washington DC

Buhyoff, G. J. and Miller, P. A. (1998) 'Context reliability and internal validity of an expert system to assess landscape visual values', *AI Applications*, vol 12, nos 1–3, pp76–82

Bull, E. L. and Johnson, E. A. (1995) 'Habitat use and management of pileated woodpeckers in northeast Oregon', *Journal of Wildlife Management*, vol 57, pp335–345

Burkey, T. V. (1989) 'Extinction in nature reserves: The effects of fragmentation and the importance of migration between reserve fragments', *Oikos*, vol 55, pp75–81

Campbell, B. and Luckert, M. (2002) *Uncovering the Hidden Harvest*, Earthscan, London

Canadian Council of Forest Ministers (1995) *Defining Sustainable Forest Management: Canadian Approach to Criteria and Indicators*, Natural Resources Canada, Ottawa, Ontario

Canadian Forest Service (undated) *Criteria and Indicators for the Conservation and Sustainable Management of Temperate and Boreal Forests – The Montreal Process*, Canadian Forest Service, Hull, Quebec

Carlson, A. (1986) 'A comparison of birds inhabiting pine plantations and indigenous forest patches in a tropical mountain area', *Biological Conservation*, vol 35, no 3

Carlson, A. (1990) 'Whose vision? Whose meanings? Whose values? Pluralism and objectivity in landscape analysis', in Groth, P. (ed.) *Vision Culture and Landscape*, working papers from the Berkeley symposium on cultural landscape interpretation, Department of Landscape Architecture, University of California, Berkeley

Carrere, R. and Lohmann, L. (1996) *Pulping the South: Industrial Tree Plantations and the World Paper Economy*, Zed Books and The World Rainforest Movement, London and Montevideo

Carter, J. (1996) *Current Approaches to Participatory Forest Resource Assessment*, Rural Development Forestry Study Guide 2, Overseas Development Institute, London

CCAD, FAO and CCAB-AP (1997) *Criteria and Indicators for Sustainable Forest Management in Central America*, CCAD, FAO and CCAB-AP, Tegucigalpa, Honduras

Chambers, R. (1997) *Whose Reality Counts? Putting the First Last*, Intermediate Technology Publications, Reading

Christiansen, M. and Hahn, K. (2003) *A Study of Dead Wood in European Beech Reserves*, NAT-MAN (Nature-based management of beech in Europe) Project

Cifuentes, M. A., Izurieta, A., Henrique De Faria, V. and H. (1999) *Medición de la Efectividad del Manejo de Areas Protegidas*, Forest Innovations Project, WWF, IUCN and GTZ, Turrialba, Costa Rica

Clark, J. (1992) 'The future of native forest logging in Australia', Working Paper 1992/1, Centre for Resource and Environmental Studies, Australian National University, Adelaide

Clausen, I. H. S. (1986) 'The use of spiders (*Aranae*) as ecological indicators', *Bulletin of the British Arachnological Society*, vol 7, no 3, pp83–86

Clifford, S. and King, A. (1993) 'Losing your place', in Clifford, S. and King, A. (eds) *Local Distinctiveness: Place, Particularity and Identity, Essays for a Conference*, Common Ground, London

Colfer, C. P., Brocklesby, M. A., Diaw, C., Etuge, P., Günter, M., Harwell, E., McDougal, C., Miyasaka, N., Porro, R., Prabhu, R., Salim, A., Sardjono, M. A., Tchinkangwa, B., Tiani, A. M., Wadley, R., Woelfel, J. and Wollenberg, E. (2000a) *The BAG (Basic Assessment Guide for Human Well-Being)*, Center for International Forestry Research, Bogor, Indonesia

Colfer, C. P., Brocklesby, M. A., Diaw, C., Etuge, P., Günter, M., Harwell, E., McDougal, C., Miyasaka, N., Porro, R., Prabhu, R., Salim, A., Sardjono, M. A., Tchinkangwa, B., Tiani, A. M., Wadley, R., Woelfel, J. and Wollenberg, E. (2000b) *The Grab Bag: Supplementary Methods for Assessing Human Well-Being*, Center for International Forestry Research, Bogor, Indonesia

Colfer, C. P., Prabhu, R., Günter, M., McDougal, C., Miyasaka, N. and Porro, R. (1999) *Who Counts Most? Assessing Human Well-Being in Sustainable Forest Management*, Center for International Forestry Research, Bogor, Indonesia

Conzen, M. P. (ed.) (1990) *The Making of the American Landscape*, Routledge, New York and London

Convention on Biological Diversity (1997a) *Indicators of Forest Biodiversity*, UNEP/CBD/Forests/LG/1/4, 12 May, paper for the meeting of the Liaison Group on Forest Biological Diversity, Helsinki, Finland, UNEP

Convention on Biological Diversity (1997b) *Recommendations for a Core Set of Indicators of Forest Biological Diversity*, UNEP/CBD/SBSTTA/3/9, 10 July, paper for the Subsidiary Body on Scientific, Technical and Technological Advice, Third Meeting, 1–5 September, UNEP, Montreal, Canada

Corrales, L. (1999) *La Necesidad de un Lenguaje Común Como Punto de Partida Proyecto: De la Teoría a la Práctica*, UICN ORMA, San José, Costa Rica

Cowling, R. M., Pressey, R. L., Sims-Castley, R., le Roux, A., Baard, E., Burgers, C. J. and Palmer, G. (2003) 'The expert or the algorithm? Comparison of priority conservation areas in the Cape Floristic Region identified by park managers and reserve selection software', *Biological Conservation*, vol 112, pp147–167

Crisp, P., Daniel, L. and Tortell, P. (1990) *Mangroves in New Zealand: Trees in the Tide*, GP Books, Auckland

Dalal-Clayton, B. and Sadler, B. (2005) *Strategic Environmental Assessment: A Sourcebook and Reference Guide to International Experience*, Earthscan, London

Danielsen, F., Balete, D. S., Poulsen, M. K., Enghoff, M., Nozawa, C. M. and Jensen, A. E. (2000) 'A simple system for monitoring biodiversity in protected areas of a developing country', *Biodiversity and Conservation*, vol 9, pp1671–1705

Davis, R. and Hirji, R. (eds) (2003a) *Environmental Flows: Concepts and Methods*, Water Resources and Environment Technical Note C.1, World Bank, Washington DC

Davis, R. and Hirji, R. (eds) (2003b) *Water Quality: Assessment and Protection*, Water Resources and Environment Technical Note D.1, World Bank, Washington DC

Davis-Case, D. (1989) *Community Forestry: Participatory Assessment, Monitoring and Evaluation*, FAO, Rome, www.fao.org/forestry/foris/data/cfu_docs/type.stm#guide

Davis-Case, D. (1990) *The Community's Toolbox: The Idea, Methods and Tools for Participatory Assessment, Monitoring and Evaluation in Community Forestry*, Community Forestry Field Manual No 2, FAO, Rome

de Beer, J. H. and McDermott, M. J. (1996) *The Economic Value of Non-Timber Forests Products in Southeast Asia*, 2nd edition, Committee for IUCN, Amsterdam, Netherlands

de Graaf, N. R. (1986) *A Silvicultural System for Natural Regeneration of Tropical Rain Forest in Suriname*, Agricultural University, Wageningen, Netherlands

Delacourt, P. A. and Delacourt, H. A. (1987) *Long-Term Forest Dynamics in the Temperate Zone*, Ecological Studies 63, Springer-Verlag, New York

Department of the Environment (1994) *Sustainable Forestry: The UK Programme*, Her Majesty's Stationery Office, London

der Steege, H. (1993) *Patterns in Tropical Rain Forest in Guyana*, Tropenbos Series 3, Tropenbos, Wageningen, Netherlands

Dorner, P. and Thiesenhusen, W. C. (1992) *Land Tenure and Deforestation: Interactions and Environmental Implications*, Discussion Paper No 34, UNRISD, United Nations, Geneva

Dudley, N. (1992) *Forests in Trouble*, WWF International, Gland, Switzerland

Dudley, N. (1995) 'Current initiatives to conserve the world's forests', *Quarterly Journal of Forestry*, vol 89, no 1, pp21–25

Dudley, N. (1996) 'Authenticity as a means of measuring forest quality', *Biodiversity Letters*, vol 3, pp6–9

Dudley, N. (1997a) 'Criteria of forest quality and forest planning at a landscape level', *Eleventh World Forestry Congress*, World Forestry Congress, Antalya, Turkey

Dudley, N. (1997b) 'Forest quality indicators', in Cohen, S. and Burgei, S. (eds) *Exploring Biodiversity Indicators and Targets under the CBD*, A Synthesis Report of the Global Biodiversity Forum, 3–4 April, Global Biodiversity Forum, New York

Dudley, N. (2003) 'L'importance de la naturalité dans les paysages forestiers', in Vallauri, D. (ed.) *Live Blanc sur la Protection des forêts Naturelles en France*, Editions TEC and DOC, Paris

Dudley, N. and Elliott, C. (1996) 'WWF Proposals for the consideration of forest quality in the Temperate and Boreal Forest Assessment TBFRA-2000', in Nyyssönen, A. and Ahti, A. (eds) *Expert Consultation on Global Forest Resources Assessment 2000: Kotka III*, Research Papers 620, Finnish Forest Research Institute, Helsinki

Dudley, N. and Jeanrenaud, J.-P. (1997) 'Needs and prospects for international cooperation in assessing forest biodiversity: An overview from WWF', in *Assessment of Biodiversity for Improved Forest Planning: Proceedings of the Conference on Assessment of Biodiversity for Improved Forest Planning, 7–11 October 1996 in Monte Verità, Switzerland*, European Forest Institute, Proceedings No 18, Kluwer Academic Publishers

Dudley, N. and Mansourian, S. (2003) *Forest Landscape Restoration and WWF's Conservation Priorities*, WWF International, Gland, Switzerland

Dudley, N. and Pollard, D. (forthcoming) *A Forest Management Tracking Tool*, WWF, Gland, Switzerland

Dudley, N. and Pressey, R. (2001) 'Forest protected areas: Why should we worry about systematic planning?', *Arborvitae Supplement*, October, WWF and IUCN, Gland, Switzerland

Dudley, N. and Rae, M. (1998) 'Criteria and indicators of forest quality', in *Proceedings of International Conference on Indicators of Sustainable Forest Management, 24–28 August, Melbourne, Australia*, IUFRO, CIFOR and FAO

Dudley, N. and Stolton, S. (1998) *A Methodology for the Rapid Assessment of Forest Quality*, first draft 25 May 1998, a working paper for the EPFL/IUCN//WWF Forest Quality Project, Equilibrium, Bristol, UK

Dudley, N. and Stolton, S. (2003a) *Running Pure: The Importance of Forest Protected Areas to Drinking Water*, WWF and World Bank, Gland, Switzerland and Washington DC

Dudley, N. and Stolton, S. (2003b) 'Ecological and socio-ecological benefits of protected areas in dealing with climate change', in Hansen, L. J., Biringer, J. L. and Hofman, J. R. (eds) *A User's Guide to Building Resistance and Resilience to Climate Change*, WWF US, Washington DC

Dudley, N. and Stolton, S. (2004) *Biological Diversity, Tree Species Composition and Environmental Protection in Regional FRA-2000*, Geneva Timber and Forest Discussion Paper 33, UNECE and FAO, New York

Dudley, N. and Stolton, S. with Gilmour, D., Jeanrenaud, J.-P., Phillips, A. and Rosabal, P. (1997) *Protected Areas for a New Millennium: The Implications of IUCN's Protected Area Categories for Forest Conservation*, WWF and IUCN, Gland, Switzerland

Dudley, N. and Vallauri, D. (2004) *Deadwood – Living Forests: The Importance of Veteran Trees and Deadwood to Biodiversity*, WWF European Programme, Brussels

Dudley, N., Barrett, M. and Baldock, D. (1985) *The Acid Rain Controversy*, Earth Resources Research, London

Dudley, N., Gilmour, D. and Jeanrenaud, J.-P. (eds) (1996) *Forests for Life*, a joint forest strategy, WWF and IUCN, Gland, Switzerland

Dudley, N., Higgins-Zogib, L. and Mansourian, S. (2006) *Beyond Belief: Linking Faiths and Protected Areas to Support Biodiversity Conservation*, WWF and ARC, Gland and Manchester

Dudley, N., Jeanrenaud, J.-P. and Sullivan, F. (1996) *Bad Harvest?*, Earthscan in association with WWF, London

Dudley, N., Nguyen Cu and Vuong Tien Manh (2003) *A Monitoring and Evaluation System for Forest Landscape Restoration in the Central Truong Son Landscape*, WWF Indochina Programme and Government of Vietnam, Hanoi

Dudley, N., Stolton, S. and Ashby, M. (1999) *A Preliminary Forest Quality Assessment of the Dyfi Catchment, Wales*, Forest Innovations Project working paper, WWF and IUCN, Gland, Switzerland

Dudley, N., Stolton, S. and Jeanrenaud, J. P. (eds) (1993) *Towards a New Definition of Forest Quality*, WWF UK, Godalming, UK

Durbin, K. (1991) 'Fishery experts say forest plans put fish at risk' *The Oregonian*, 29 October, Portland

Dyson, M., Bergkamp, G. and Scanlon, J. (eds) (2003) *Flow: The Essentials of Environmental Flow*, IUCN, Gland, Switzerland

Eeronheimo, O., Ahti, A. and Sahlberg, S. (1997) *Criteria and Indicators for Sustainable Forest Management in Finland*, Ministry of Agriculture and Forestry, Helsinki

Ekelund, H. and Dahlin, C.-G. (1997) *Development of the Swedish Forests and Forest Policy During the Last 100 years*, Skogsstyrelsen, Jönköping

Elliott, C. (1995) *A WWF Guide to Forest Certification*, WWF UK, Godalming

Elosegi, A., Diez, J. R. and Pozo, J. (1999) 'Abundance, characteristics, and movement of woody debris in four Basque streams', *Archiv Fur Hydrobiologie*, vol 144, pp455–471

Emerson, R. W. (1836) *Nature and Other Writings*, (edited by Peter Turner) Shambhala Publications, Boston

Emerton, L. (2001) 'Why forest values are important to East Africa', *Innovations*, vol 8, no 2, African Centre for Technology Studies, Nairobi

Emrich, A., Pokormy, B. and Sepp, C. (2000) *The Significance of Secondary Forest Management for Development Policy*, GTZ, Eschborn

Ervin, J. (2003) *WWF Rapid Assessment and Prioritization of Protected Area Management (RAPPAM) Methodology*, WWF, Gland, Switzerland

Esseen, P. A. (1994) 'Tree mortality patterns after experimental fragmentation of an old-growth conifer forest', *Conservation Biology*, vol 68, pp19–28

Fairhead, J. and Leach, M. (1996) *Misreading the Africa Landscape*, Cambridge University Press, Cambridge

FAO (United Nations Food and Agriculture Organization) (1980) *Impacts on Soils of Fast-Growing Species in Lowland Humid Tropics*, FAO Forestry Paper no 21, FAO, Rome

FAO (1986) *Forest Policies in Europe*, FAO Forestry Paper 86, FAO, Rome

FAO (1993) *Conservation of Genetic Resources in Tropical Forest Management: Principles and Concepts*, FAO Forestry Paper no 107, FAO, Rome

FAO (1995) *Non-Wood Forest Products for Rural Income and Sustainable Forestry*, Non-Wood Forest Products 7, FAO, Rome

FAO (1996) *Criteria and Indicators for Sustainable Forest Management in Dry-Zone Africa: UNEP/FAO Expert Meeting, Nairobi, Kenya, 21–24 November 1995*, FAO, Rome

FAO (1997) *State of the World's Forests 1997*, FAO, Rome

FAO and CIFOR (Center for International Forestry Research) (2005) *Forests and Floods: Drowning in Fiction or Thriving on Facts?* FAO and CIFOR, Bangkok and Bogor, Indonesia

FAO and UNECE (2000) *Forest Resources of Europe, CIS, North America, Australia, Japan and New Zealand: Main Report*, FAO and UNECE, Geneva and Rome

Ferreira, L. V., Lemos de Sá, R. M., Buschbacher, R., Batmanian, G., Bensusan, N. R. and Costa, K. L. (1999) 'Protected areas or endangered spaces?', in Barbosa, A. C. and Lacava, U. (eds) *WWF Report on the Degree of Implementation and the Vulnerability of Brazilian Federal Conservation Areas*, WWF Brazil, Brasília

Ferris-Kaan, R. (1991) *Edge Management in Woodlands*, Occasional Paper 28, Forestry Commission, Edinburgh

Ferris-Kaan, R., Peace, A. J. and Humphrey, J. W. (1996) 'Assessing structural diversity in managed forests', *Forestry Sciences*, vol 51, pp331–342

Fisher, R. J. (1994) 'Indigenous forest management in Nepal: Why common property is not a problem', in Allen, M. (ed.) *Anthropology of Nepal: People, Problem and Processes*, Mandala Book Point, Kathmandu

Fisher, R. J., Maginnis, S., Jackson, W. J., Barrow, E. and Jeanrenaud, S. (2005) *Poverty and Conservation: Landscapes, People and Power*, IUCN, Gland, Switzerland

Flannery, T. (1994) *The Future Eaters*, Reed New Holland, Sydney

Florio, M. (1992) 'Italy', in Wibe, S., and Jones, T. (eds) *Forests: Market and Intervention Failures – Five Case Studies*, Earthscan, London

Ford, E. D., Malcolm, D. C. and Atterson, J. (eds) (1979) *The Ecology of Even-Aged Forest Plantations*, Institute of Terrestrial Ecology, Cambridge

Franklin, J. F., Cromack Jr, K., Denison, W., McKee, A., Maser, C., Sedell, J., Swanson, F. and Juday, G. (1981) *Ecological Characteristics of Old-Growth Douglas-Fir Forests*, General Technical Report PNW-118, Pacific Northwest Forest and Range Experimental Station, USDA Forest Service, Portland, Oregon

Freemark, K. E. and Merrimen, H. G. (1986) 'Importance of area and habitat heterogeneity to bird assemblages in temperate forest fragments', *Biological Conservation*, vol 36, pp115–141

Freese, C. H. (1997) *Harvesting Wild Species: Implications for Biodiversity Conservation*, John Hopkins University Press, Baltimore and London

Fridman, J. and Walheim, M. (2000) 'Amount, structure and dynamics of dead wood on managed forest land in Sweden', *Forest Ecology and Management*, vol 131, pp23–36

Garforth, M. and Dudley, N. (2003) *Forest Renaissance*, WWF UK and the UK Forestry Commission, Godalming and Edinburgh

Global Forest Watch (2000) *An Overview of Logging in Cameroon*, World Resources Institute, Washington DC

Godoy, R. A., Lubowski, R. and Markandaya, A. (1993) 'A method for the economic valuation of non-timber tropical forest products', *Economic Botany*, vol 47, no 3, pp220–233

Gomez-Pompa, A., Whitmore, T. C. and Hadley, M. (eds) (1991) *Rainforest Regeneration and Management*, MAB and UNESCO, Parthenon Publishing, Paris

Grace, J., Lloyd, J., McIntyre, J., Miranda, A. C., Meir, P., Miranda, H. S., Nobre, C., Moncrieff, J., Massheder, J., Malhi, Y., Wright, I. and Gash, J. (1995) 'Carbon dioxide uptake by an undisturbed tropical rain-forest in Southwest Amazonia, 1992–1993', *Science*, vol 270, pp778–780

Grant, G. (1990) 'Hydrologic, geomorphic and aquatic habitat implications of old and new forestry', paper presented at *Forests – Wild and Managed: Differences and Consequences, 19–20 January, Vancouver*, University of British Columbia, Vancouver, British Columbia

Green, I. P., Higgins, R. J., Kitchen, C. and Kitchen, M. A. R. (2000) *The Flora of the Bristol Region*, Wildlife of the Bristol Region 1, Pisces Publications and Bristol Regional Environmental Records Centre, Newbury and Bristol

Groombridge, B., and Jenkins, M. D. (1996) 'Assessing biodiversity: Status and sustainability', in *WCMC Biodiversity*, vol 5, World Conservation Press, Cambridge

Gustafsson, L. and Hallingbäck, T. (1988) 'Bryophyte flora and vegetation of managed and virgin coniferous forests in southwest Sweden', *Biological Conservation*, vol 44, pp283–300

Hamilton, E. H. and Yearsley, H. K. (1988) *Vegetation Development after Clearcutting and Site Preparation in the Sub-Boreal Spruce Zone*, FRDA Report 018, FRDA, British Columbia, Canada

Hamilton, L. S. (1987) 'What are the impacts of Himalayan deforestation on the Ganges-Brahamaputra lowlands and delta? Assumptions and facts', *Mountain Research and Development*, vol 7, no 3, pp256–263

Hammond, A., Adriaanse, A., Rodenburg, E., Bryant, D. and Woodward, R. (1995) *Environmental Indicators: A Systematic Approach to Measuring and Reporting on Environmental Policy Performance in the Context of Sustainable Development*, World Resources Institute, Washington DC

Harper, J. L. and Hawksworth, D. L. (eds) (1995) 'Biodiversity measurement and estimation', *Philosophical Transactions of the Royal Society of London B*, vol 345, pp89–99

Harris, L. D. (1984) *The Fragmented Forest: Island Biogeography Theory and Preservation of Biotic Diversity*, University of Chicago Press, Chicago

Hartzell Jr, H. (1991) *The Yew Tree: A Thousand Whispers*, Hulogosi, Eugene, Oregon

Hausser, J. (1995) 'Säugetieratlas der Schweiz', *Denkschriftenkommission der Schweizerischen Akademie der Naturwissenschaften*, Birkhäuser Verlag, Basel

Hawksworth, D. L. (ed) (1996) *Biodiversity*, CAB Books, Oxford

Hawksworth, D. L. and Rose, F. (1976) *Lichens as Pollution Monitors*, Edward Arnold in association with the Institute of Biology, London

Hawksworth, D. L., Kirk, P. M. and Dextre Clarke, S. (eds) (1997) *Biodiversity Information Needs and Options*, CAB International, Oxford

Henriksen, A., Kamari, J., Posch, M. and Wilander, A. (1992) 'Critical loads of acidity: Nordic surface waters', *Ambio*, vol 21, pp356–363

Herrera, B. and Corrales, L. (1999) *Propuesta Metodológica Para la Selección de Criterios e Indicadores y Análisis de Verificadores Relativos a Calidad de Bosque y a Nivel de Paisaje*, UICN ORMA, San José, Costa Rica

Herrera, B. and Salas, A. (1999a) 'Estándares para la evaluación y el monitoreo de la calidad del bosque a nivel de paisaje', UICN ORMA, San José, Costa Rica

Herrara, B. and Salas, A. (1999b) *Protocol for Gathering and Recording Information for the Evaluation and Monitoring of Forest Quality at a Landscape Level*, IUCN ORMA, San José, Costa Rica

Herrara, B. and Salas, A. (1999c) *Standards for Evaluating and Monitoring Forest Quality at a Landscape Level: Final Document*, IUCN ORMA, San José, Costa Rica

Herrara, B. and Salas, A. (1999d) *Standards for Evaluating and Monitoring Forest Quality at a Landscape Level: Training Module*, IUCN ORMA, San José, Costa Rica

Hockings, M., Stolton, S. and Dudley, N. (2000) *Evaluating Effectiveness: A Framework for Assessing the Management of Protected Areas*, Cardiff and Cambridge, Cardiff University and IUCN

Hoekstra, J. M., Boucher, T. M., Ricketts, T. H. and Roberts, C. (2005) 'Confronting a biome crisis: Global disparities of habitat loss and protection', *Ecology Letters*, vol 8, pp23–29

Hornback, K. E. and Eagles, P. F. J. (1999) *Guidelines for Public Use Measurement and Reporting at Parks and Protected Areas*, Cooperative Research Centre for Sustainable Tourism of Australia, IUCN and Parks Canada, Gland, Switzerland

Horneman, L. N., Beeton, R. J. S. and Hockings, M. (2002) *Monitoring Visitors to Natural Areas: A Manual on Standard Methodological Guidelines*, University of Queensland, Gatton Campus, Australia

Hulme, M. and Viver, D. (1998) 'A climate change scenario for the tropics', *Climate Change*, vol 39, pp145–176

Humphreys, D. (1996) *Forest Politics: The Evolution of International Co-operation*, Earthscan, London

Iacobelli, T., Kavanagh, K. and Rowe, S. (1994) *A Protected Areas Gap Analysis Methodology: Planning for the Conservation of Biodiversity*, WWF Canada, Toronto

IAITPTF and IWGIA (undated) *Indigenous Peoples, Forest and Biodiversity*, International Alliance of Indigenous-Tribal Peoples of the Tropical Forest and International Work Group for Indigenous Affairs, Copenhagen and London

Ingles, A. W. (1995) 'Religious beliefs and rituals in Nepal: Their influence on forest conservation', in Halladay, P. and Gilmour, D. A. (eds) *Conserving Biodiversity Outside Protected Areas: The Role of Traditional Agro-Ecosystems*, IUCN, AMA-Andalucía and Centro de Investigacíon F. González-Bernáldez, Gland, Switzerland

Iorgulescu, I. (1997) *Quantitative Analysis of Landscape Fragmentation*, EPFL, Lausanne

ITTC (1997) *Report of the Expert Panel on Criteria and Indicators for the Measurement of Sustainable Management in Natural Tropical Forests, Yokohama, 8–12 September*, Document ITTC(XXIII)/9, 8 October, International Tropical Timber Council

ITTO (undated) *ITTO Guidelines on the Conservation of Biological Diversity in Tropical Production Forests*, ITTO Policy Development Series No 5, International Tropical Timber Organization, Yokahama, Japan

ITTO (1992) *Criteria for the Measurement of Sustainable Tropical Forest Management*, ITTO, Policy Development Series No 3, International Tropical Timber Organization, Yokahama, Japan

ITTO (1993) *ITTO Guidelines for the Establishment and Sustainable Management of Planted Tropical Forests*, International Tropical Timber Organization, Yokahama, Japan

ITTO (2002) *ITTO Guidelines for the Restoration, Management and Rehabilitation of Degraded and Secondary Tropical Forests*, International Tropical Timber Organization, Yokahama, Japan

ITTO (2003) *ATO / ITTO Principles, Criteria and Indicators for the Sustainable Management of African Natural Tropical Forests*, ITTO Policy Development Series No 14, International Tropical Timber Organization, Yokahama, Japan

IUCN (2000) *IUCN Red List Categories and Criteria*, IUCN Species Survival Commission, 51st Meeting of the IUCN Council, IUCN, Gland, Switzerland

IUCN and WWF (1996) *Criteria and Indicators of Forest Quality: A Paper for the IPF*, discussion paper for the Third Meeting of the Intergovernmental Panel on Forests of the Commission on Sustainable Development, 9–20 September, Geneva, IUCN and WWF

Jackson, W. J. and Ingles, A. (1998) *Participatory Techniques in Community Forestry*, IUCN, Gland, Switzerland

Jackson W. J., Jeanrenaud, J.-P. and Dudley, N. (2000) *Forests for Life: Reaffirming the Vision*, IUCN and WWF, Gland, Switzerland

Janisch, J. E. and Harmon, M. E. (2002) 'Successional changes in live and dead wood carbon stores: Implications for ecosystem productivity', *Tree Physiology*, vol 22, pp77–89

Jennings, S. and Jarvie, J. (with input from Dudley, N. and Deddy, K.) (2003) *A Sourcebook for Landscape Analysis of High Conservation Value Forests* (Version 1), ProForest, Oxford

Jennings, S., Nussbaum, R., Judd, N. and Evans, T. (2003) *The High Conservation Value Forest Toolkit*, three part report, ProForest, Oxford

Johns, A. D. (1996) 'Bird population persistence in Sabahan logging concessions', *Conservation Biology*, vol 75, no 1, pp3–10

Johnson, K. N., Franklin, J. F., Thomas, J. W. and Gordon, J. (1991) *Alternatives to Late-Successional Forests of the Pacific Northwest*, a Report to the Agriculture Committee and The Merchant Marine and Fisheries Committee of the US House of Representatives by the Scientific Panel on Late-Successional Forest Ecosystems, 8 October

Johnson, N., White, A. and Perot-Maitre, D. (2001) *Developing Markets for Water Services from Forests*, Forest Trends, World Resources Institute and The Katoomba Group, Washington DC

Kandler, O. (1992) 'Historical declines and diebacks of central European forests and present conditions', *Environmental Toxicology and Chemistry*, vol 11, pp1077–1093

Kapos, V., Blyth, S., Fritz, S., Flamenco, A., Iremonger, S., Aldrich, M. and Quinton, T. (1997) *Developing Indicators of the State of the World's Tropical Forests*, draft, World Conservation Monitoring Centre for WWF, Cambridge

Kapos, V., Jenkins, M. D., Lysenko, I., Ravilious, C., Bystriakova, N. and Newton, A. (2001) *Forest Biodiversity Indicators*, UNEP-World Conservation Monitoring Centre, Cambridge

Karr, R. J. (1994) 'Ecological integrity and ecological health are not the same', in Schulze, P., Frosch, R. and Risser, P. (eds) *Engineering Within Ecological Constraints*, National Academy of Engineering, Washington DC

Karström, M., Lindahl, K., Olsson, G. A. and Williamson, M. (1993) *Indikatorarter för Identifiering av Naturskogar i Norrbotten*, Naturvårdsverket, Stockholm

Kirby, K. J. (1988) *A Woodland Survey Handbook*, Research and Survey in Nature Conservation Series Series 11, Nature Conservancy Council, Peterborough, UK

Kirby, K. J. (1990) 'Changes in the ground flora of a broad leaved wood within a clear fell, group fell and a coppiced block', *Forestry*, vol 63, pp242–249

Kirby, K. J. (1992a) 'Accumulation of dead wood: A missing ingredient in coppicing?', in Buckley, G. P. (ed.) *Ecology and Management of Coppice Woodlands*, Chapman and Hall, London

Kirby, K. J. (1992b) *Woodland and Wildlife*, Whittet Books, London

Kohm, K. A. and Franklin, J. F. (eds) (1997) *Creating a Forestry for the 21st Century: The science of ecosystem management*, Island Press, Washington DC and Covelo, California

Kothari, A., Pande, P., Singh, S. and Variava, D. (1989) *Management of National Parks and Sanctuaries in India, A Status Report*, Environmental Studies Division, Indian Institute of Public Administration, New Delhi

Küchli, C. (1997) *Forests of Hope: Stories of Regeneration*, Earthscan, London

Küchli, C., Bollinger, M. and Rüsch, W. (1998) *The Swiss Forest – Taking Stock: Interpretation of the Second National Forest Inventory in Terms of Forestry Policy*, Swiss Agency for the Environment, Forests and Landscape, Bern

Kuusipalo, J. (1984) 'Diversity pattern of the forest understorey vegetation in relation to some site characteristics', *Silva Fennica*, vol 18, no 2, pp121–131

Laird, S. A. (ed.) (2002) *Biodiversity and Traditional Knowledge: Equitable Partnerships in Practice*, WWF and Earthscan, London

Langford, K. J. (1976) 'Change in yield of water following a bushfire in a forest of *Eucalyptus reglans*', *Journal of Hydrology*, vol 89, pp87–114

Leach, G. and Mearns, R. (1988) *Beyond the Woodfuel Crisis: People, Land and Trees in Africa*, Earthscan, London

Lee, C. and Schaaf, T. (eds) (2003) *The Importance of Sacred Natural Sites for Biodiversity Conservation*, UNESCO, Paris

Linden, O. (1990) 'Human impact on tropical coastal zones', *Nature and Resources*, vol 26, no 4, pp3–11

Loh, J. (2003) *Living Planet Index 2003*, WWF and the UNEP-World Conservation Monitoring Centre, Gland, Switzerland and Cambridge, UK

Loikkanen, T., Simojoki, T. and Wallenius, P. (1999) *Participatory Approach to Natural Resource Management*, Metsähallitus Forest and Park Service, Vantaa, Finland, www.metsa.fi/page.asp?Section=1200&Item=1644

Lorimer, C. G. (1989) 'Relative effects of small and large disturbances on temperate hardwood forest structure', *Ecology*, vol 70, pp565–575

Loucks, C., Springer, J., Palmineri, S., Morrison, J. and Strand, H. (2004) *From the Vision to the Ground: A Guide to Implementing Ecoregion Conservation in Priority Areas*, WWF, Washington DC

Luoma, J. R. (1999) *The Hidden Forest: The Biography of an Ecosystem*, Owl Books, New York

Mader, H. J., Schell, C. and Kornacker, P. (1990) 'Linear barriers to movements in the landscape', *Biological Conservation*, vol 54, pp209–222

Magill, A. W. and Schwartz, C. F. (1989) *Searching for the Value of a View*, Research paper PSW-193, Pacific Southwest Forest and Range Experimental Station, Forest Service, US Department of Agriculture, Berkeley, California

Maginnis, S., Jackson, W. J. and Dudley, N. (2004) 'Conservation landscapes: Whose landscapes? Whose tradeoffs?', in McShane, T. and Wells, M. (eds) *Getting Biodiversity Projects to Work*, Columbia University Press, New York

Mansourian, S., Valauri, D. and Dudley, N. (eds) (2005) *Forest Restoration in Landscapes: Beyond Planting Trees*, Springer, New York

Markham, A., Dudley, N. and Stolton, S. (1993) *Some Like It Hot*, WWF International, Gland, Switzerland

Martin, V. (ed.) (1982) *Wilderness*, Findhorn Press, Findhorn, Scotland

Martin, W. H. (1992) 'Characteristics of old-growth mixed mesophytic forests', *Natural Areas Journal*, vol 12, pp127–135

Maser, C., Trappe, J. M. and Franklin, J. F. (eds) (1988) *From the Forest to the Sea: A Story of Fallen Trees*, General Technical Report PNW-GTR-229, USDA Forest Service, Pacific Northwest Research Station, Portland, Oregon

Matthews, E., Payne, R., Rohweder, M. and Murray, S. (2000) *Pilot Analysis of Global Ecosystems: Forest Ecosystems*, World Resources Institute, Washington DC

McFarlane, D. (1991) 'A review of secondary salinity in agricultural areas of Western Australia', *Land and Water Research News*, no 11, pp7–166

McCormick, N. and Folving, S. (1998) 'Monitoring European forest biodiversity at regional scales using remote sensing', *Forestry Sciences*, vol 51, pp283–290

MCPFE (1995) *European Criteria and Indicators for Sustainable Forest Management: Adopted by the Expert Level Follow-Up Meetings of the Helsinki Conference in Geneva, 24 June 1994 and in Antalya, 23 January 1995*, Ministerial Conference for the Protection of Forests in Europe

MCPFE (2002) *Improved Pan-European Indicators for Sustainable Forest Management adopted by the MCPFE Expert-level meeting, 7–8 October, Vienna*, Ministerial Conference on the Protection of Forests in Europe, Vienna

MCPFE Liaison Unit and FAO (2003) *State of Europe's Forests 2003: The MCPFE Report on Sustainable Forest Management in Europe*, Ministerial Conference on the Protection of Forests in Europe, Vienna

Messerschmidt, D. A. (ed.) (1993) *Common Forest Resource Management: Annotated Bibliography of Asia, Africa and Latin America*, Community Forestry Note no 11, FAO, Rome

Messerschmidt, D. A. (1995) *Rapid Appraisal for Community Forestry: The RA Process and Rapid Diagnostic Tools*, International Institute for Environment and Development, London

Michon, G. and de Foresta, H. (1995) 'The Indonesian agro-forest model', in Halladay, P. and Gilmour, D. A. (eds) *Conserving Biodiversity Outside Protected Areas: The Role of Traditional Agro-ecosystems*,

IUCN in collaboration with AMA Andalucia and Centro de Investigación F González-Bernáldez, Gland, Switzerland

Miller, K. and Lanou, S. (1995) *National Biodiversity Planning*, World Resources Institute, Washington DC

Miller, S. A., Maloney, K. O. and Feminella, J. W. (2004) 'When and why is coarse woody debris a refuge for biofilm in sandy coastal plains streams?', in *North American Benthological Society Annual Meeting, Vancouver, British Columbia*, North American Benthological Society

Ministère de l'agriculture et de la pêche (1994) *Indicators for the Sustainable Management of French Forests*, Ministère de l'agriculture et de la pêche, Paris

Ministry of Environment (2002) *Living with Nature: The National Biodiversity Strategy of Japan*, Nature Conservation Bureau, Ministry of Environment, Tokyo

Ministry of Foreign Affairs of Peru (1995) *Regional Workshop on the Definition of Criteria and Indicators for Sustainability of Amazonian Forests: Final Document*, February 25, Ministry of Foreign Affairs of Peru, Tarapoto, Peru

Mitchell, J. H. (2001) *The Wildest Place: Italian Gardens and the Invention of Wilderness*, Counterpoint, Washington DC

Mittermeier, R. A., Goettsch Mittermeier, C., Robles Gil, P. and Pilgrim, J. (2003) *Wilderness: Earth's Last Wild Places*, University of Chicago Press, Chicago

Mladenoff, D. J., White, M. A., Pastor, J. and Crow, T. (1993) 'Comparing spatial patterns in unadulterated old-growth and disturbed forest landscapes', *Ecological Applications*, vol 3, pp294–306

Moiseev, A. with Dudley, E. and Cantin, D. (2002) *The Wellbeing of Forests – An e-Tool for Assessing Environmental and Social Sustainability*, IUCN, Temperate and Boreal Forest Programme, IUCN Forest Conservation Programme, Gland, Switzerland

Molnar, A., Scherr, S. and Khare, A. (2004) *Who Conserves the World Forests? A New Assessment of Conservation and Investment Trends*, Forest Futures and Ecoagriculture Partners, Washington DC

Morris, W., Doak, D., Groom, M., Kareiva, P., Fieberg, J., Gerber, L., Murphy, P. and Thompson, D. (1999) *A Practical Handbook for Population Viability Analysis*, The Nature Conservancy, Arlington, Virginia

Mount, A. B. (1973) 'Natural regeneration processes in Tasmanian forests', *Search*, vol 10, pp180–186

Moussouris, Y. and Regatto, P. (1999) *Forest Harvest: An Overview of Non Timber Forest Products in the Mediterranean Region*, WWF Mediterranean Programme, Rome

Nature Conservancy, The (2000) *The Five-S Framework for Site Conservation*, two volumes, The Nature Conservancy, Arlington, Virginia

Negussie, G. (1997) *Use of Traditional Values in the Search for Conservation Goals: The Kaya Forests of the Kenyan Coast*, paper presented at African Rainforests and the Conservation of Biodiversity Conference, Limbe Botanic Gardens, South West Cameroon, 17–24 January

Nehlsen, W., Williams, J. E. and Lichatowich, J. A. (1991) 'Pacific salmon at the crossroads: Stocks at risk from California, Oregon, Idaho and Washington', *Fisheries*, vol 16, no 2, pp4–21

Nemarundwe, N., de Jong, W. and Cronkleton, P. (2003) *Future Scenarios as an Instrument for Forest Management: Manual for Training Facilitators of Future Scenarios*, Center for International Forestry Research, Bogor, Indonesia, www.cifor.cgiar.org/scripts/newscripts/publications/default.asp

Nilsson, S. (1996) *Do We Have Enough Forests?*, Occasional Paper no 5, International Union of Forest Research Organisations, Vienna

Norton-Griffiths, M. (1979) 'The influence of grazing, browsing and fire on the vegetation dynamics of Serengeti', in Sinclair, A. R. E. and Norton-Griffiths, M. (eds) *Serengeti: Dynamics of an Ecosystem*, University of Chicago Press, Chicago, pp310–352

Noss, R. (2001) 'Beyond Kyoto: Forest management in a time of rapid climate change', *Conservation Biology*, vol 15, no 3, pp578–590

Nugroho, T. and Siliew, I. (1997) 'Tree gardens in Sumatra', paper for the Expert Panel on Trade and Sustainable Development, unpublished

Nyyssönen, A. and Ahti, A. (eds) (1996) *Expert Consultation on Global Forest Resources Assessment 2000: Kotka III*, Research Papers 620, The Finnish Forest Research Institute, Helsinki

Ogden, C. L. (1991) *Guidelines for Integrating Nutrition Concerns into Forestry Projects*, Community Forestry Field Manual, FAO, Rome

Oliver, C. D. and Larson, B. C. (1990) *Forest Stand Dynamics*, McGraw-Hill, New York

Olsen, D. M., Dinerstein, E., Wilkramanayake, E. D., Burgess, N. D., Powell, G. V. N., Underwood, E. C., d'Amico, J. A., Itoula, I., Strand, H. E., Morrison, J. C., Louks, C. J., Allnutt, T. F., Ricketts, T. H., Kura, Y., Lamoreux, J. F., Wettengel, W. W., Hedao, P. and Kassem, K. R. (2001) 'Terrestrial ecoregions of the world: A new map of life on Earth', *Bioscience*, vol 51, no 11, pp933–938

Pagiola, S., Bishop, J. and Landell-Mills, N. (eds) (2002) *Selling Forest Environmental Services: Market-based Mechanisms for Conservation and Development*, Earthscan, London

Pakenham, T. (1996) *Meetings with Remarkable Trees*, Weidenfeld and Nicholson, London

Pakenham, T. (2002) *Remarkable Trees of the World*, Weidenfeld and Nicholson, London

Park, G. (1995) *Ng Uruora: The Groves of Life – Ecology and History in a New Zealand Landscape*, Victoria University Press, Wellington

Parvianen, J., Kassioumis, K., Bückling, W., Hochbichler, E., Päivinen, R. and Little, D. (2000) *Forest Reserves Research Network in Europe, COST Action E4: Mission, Goals, Outputs, Linkages, Recommendations and Partners: Final Report*, The Finnish Forest Research Station, Joensuu

Pastor, J. and Broschart, M. (1990) 'The spatial pattern of a northern-conifer hardwood landscape', *Landscape Ecology*, vol 4, pp55–68

Perry, D. A. (1993) 'Biodiversity and wildlife are not synonymous', *Conservation Biology*, vol 7, no 1, pp204–205

Peterken, G. (1998) *Naturalness*, Cambridge University Press, Cambridge

Peterken, G. (2002) *Reversing the Habitat Fragmentation of British Woodlands*, WWF UK, Godalming

Phillips, S. J. and Wentworth Comus, P. (eds) (2000) *The Natural History of the Sonoran Desert*, Arizona-Sonora Desert Museum, Tuscon

Picket, S. T. A. and White, P. S. (1985) *The Ecology of Natural Disturbance and Patch Dynamics*, Academic Press, New York

Pimbert, M. and Pretty, J. (1995) *Parks, People and Professionals: Putting 'Participation' into Protected Area Management*, UNRISD Discussion Paper No 57, UN Research Institute for Social Development, Geneva

Porteous, A. (1928) *Forest Folklore*, George Allen & Unwin, London

Posey, D. A. and Balee, W. (eds) (1989) *Resource Management in Amazonia: Indigenous and Folk Strategies*, The New York Botanical Garden, New York

Poulsen, J., Applegate, G. and Raymond, D. (2001) *Linking C&I to a Code of Practice for Industrial Tropical Tree Plantations*, Center for International Forestry Research, Bogor, Indonesia

Prabhu, R., Colfer, C. J. P., Venkateswarlu, P., Lay Cheng Tan, Soekmadi, R. and Wollenberg, E. (1996) *Testing Criteria and Indicators for the Sustainable Management of Forests: Phase 1 Final Report*, Center for International Forestry Research, Bogor, Indonesia

Prabhu, R., Colfer, C. J. P. and Dudley, R. G. (1999) *Guidelines for Developing, Testing and Selecting Criteria and Indicators for Sustainable Forest Management*, Criteria and Indicator Toolbox Series No 1, Center for International Forestry Research, Bogor, Indonesia

Prescott-Allen, R. (2001) *The Well-Being of Nations: A Country-by-Country Index of Quality of Life and the Environment*, Island Press, Covelo, California and Washington DC

Pyle, S. J. (1997) *Vestal Fire: An Environmental History, Told through Fire, of Europe and Europe's Encounter with the World*, University of Washington Press, Seattle and London

Rackham, O. (1996) *Trees and Woodlands in the British Landscape*, Weidenfeld and Nicholson, London

Ranney, J. W., Bruner, M. C. and Levenson, J. B. (1981) 'The importance of edge in the structure and dynamics of forest islands', in Burgess, R. L. and Sharpe, D. M. (eds) *Forest Island Dynamics in Man-Dominated Landscapes*, Springer-Verlag, New York

Raphael, M. G. and White, M. (1984) 'Use of snags by cavity-nesting birds in the Sierra Nevada', *Wildlife Monographs*, vol 86, pp1–66

Ratcliffe, P. R. (1993) *Biodiversity in Britain's Forests*, The Forestry Authority, Forestry Commission, Edinburgh

Reid, W. V., McNeely, J. A., Tunstall, D. B., Bryant, D. A. and Winograd, M. (1993) *Biodiversity Indicators for Policy-Makers*, WRI and IUCN, Washington DC and Gland, Switzerland

Repenning, R. W. and Labisky, R. F. (1985) 'Effects of even-aged timber management on bird communities of the longleaf pine forest in northern Florida', *Journal of Wildlife Management*, vol 49, pp1088–1098

Ripple, W. J., Bradshaw, G. A. and Spies, T. A. (1991) 'Measuring forest landscape patterns in the Cascade range of Oregon, USA', *Biological Conservation*, vol 57, pp73–88

Ritchie, B., McDougall, C., Haggith, M. and Burford de Oliveira, N. (2000) *Criteria and Indicators of Sustainability in Community Managed Forest Landscapes*, Center for International Forestry Research and others, Bogor, Indonesia

Rojas, M. and Aylward, B. (2002) *Cooperation Between a Small Private Hydropower Producer and a Conservation NGO for Forest Protection: The Case of La Esperanza, Costa Rica*, Land-Water Linkages in Rural Watersheds Case Study Series, FAO, Rome

Rolstad, J. and Wegge, P. (1987) 'Distribution and size of capercaillie leks in relationship to old forest fragmentation', *Oecologia*, vol 72, pp389–394

Ruitenbeek, H. J. (1990) *Economic Analysis of Conservation Initiatives: Examples from West Africa*, WWF UK, Godalming

Ruiz Pérez, M. and Arnold, J. E. M. (1996) *Current Issues in Non-Timber Forest Products Research*, CIFOR and ODA, Bogor, Indonesia

Runkle, J. R. (1982) 'Patterns of disturbance in some old-growth mesic forests of eastern North America', *Ecology*, vol 62, pp1041–1546

Saatchi, S., Agosti, D., Alger, K., Delabie, J. and Musinsky, J. (2001) 'Examining fragmentation and loss of primary forest in the southern Bahian Atlantic Forest of Brazil with radar imagery', *Conservation Biology*, vol 15, no 4, pp867–875

Safford, L. and Maltby, E. (eds) (1998) *Guidelines for Integrated Planning and Management of Tropical Lowland Peatlands: With Special Reference to Southeast Asia*, IUCN Commission on Ecosystem Management Tropical Peatland Expert Group, IUCN, Gland, Switzerland

Salim, A., Colfer, C. J. P. and McDougall, C. (2000) *Scoring and Analysis Guide*, Center for International Forestry Research, Bogor, Indonesia

Sallabanks, R., Arnett, E., Bentley Wigley, T. and Irwin, L. (2001) *Accommodating Birds in Managed Forests of North America: A Review of Bird–Forestry Relationships*, Technical Bulletin No 822, National Council for Air and Stream Improvement, Research Triangle Park, Raleigh

Sanderson, E., Redford, K. H., Vedder, A., Ward, S. E. and Coppolillo, P. B. (2002) 'A conceptual model for conservation planning based on landscape species requirements', *Landscape and Urban Planning*, vol 58, pp41–56

Sandström, E. (2003) 'Dead wood: Objectives, results and life-projects in Swedish forestry', in Mason, F., Nardi, G. and Tisato, M. (eds) *Legno Morto: Una Chiave per a Biodiversita / Dead Wood: A Key to Biodiversity*, Proceedings of the International Symposium, 29–31 May, Mantova, Italy, Sherwood 95 Supplement 2

Sayer, J. and Ruiz Pérez, M. (1994) 'What do non-timber products mean for forest conservation?', *IUCN The World Conservation Union Bulletin*, vol 25, no 3

Sayre, R., Roca, E., Sedaghatkish, G., Young, B., Keel, S., Roca, R. and Sheppard, S. (2000) *Nature in Focus: Rapid Ecological Assessment*, The Nature Conservancy and Island Press, Washington DC and Covelo, California

Schmid, H., Luder, R., Naef-Daenzer, B., Graf, R. and Zbinden, N. (1998) *Schweizer Brutvogelatlas Atlas des Oiseaux Nicheurs de Suisse. Verbreitung der Brutvogel in der Schweiz and im Furstentum*

Liechtenstein 1993–1996. Distribution des Oiseaux Nicheurs en Suisse et au Liechtenstein en 1993–1996, Schweizerische Vogelwarte Sempach, Switzerland

Schoonmaker Freudenberger, K. (1994) *Tree and Land Tenure: Rapid Appraisal Tools*, Community Forestry Field Manual No 4, Rome, FAO, www.fao.org/forestry/foris/data/cfu_docs/type.stm#guide

Schütt, P., Kock, W., Blaschke, H., Lang, K. J., Schuck, H. J. and Summerer, H. (1983) *So Stirbt der Wald*, Munich

Scott, M. J., Davis, F. W., Cusuti, B., Noss, R., Butterfield, B., Groves, C., Anderson, H., Caicco, S., D'Erchia, F., Edwards, T. C., Ulliman, J. and Wright, R. G. (1993) 'GAP analysis: A geographic approach to protection of biological diversity', *Wildlife Monographs*, vol 123, pp1–41

Sedaghatkish, G. (1999) *Rapid Ecological Assessment Sourcebook*, The Nature Conservancy, Arlington, Virginia

Sedjo, R. A. (1999) 'The potential of high-yield plantation forestry for meeting timber needs', *New Forests*, vol 17, pp339–359

Sheil, D., Puri, R. K., Basuki, I., van Heist, M., Wan, M., Liswanti, N., Rukmiyati, Sardjono, M. A., Samsoedin, I., Sidiyasa, K. D., Chrisandini, Permana, E., Angi, E. M., Gatzweiler, F., Johnson, B. and Wijaya, A. (2003) *Exploring Biological Diversity, Environment and Local People's Perspectives in Forest Landscapes: Methods for a Multidisciplinary Landscape Assessment*, Center for International Forestry Research, Bogor, Indonesia

Shepheard, P. (1997) *The Cultivated Wilderness: Or What is landscape?*, MIT Press, Cambridge

Sithole, B. (2002) *Where the Power Lies: Multiple Stakeholder Politics Over Natural Resources – A Participatory Methods Guide*, Center for International Forestry Research, Bogor, Indonesia, www.cifor.cgiar.org/scripts/newscripts/publications/default.asp

Smith, M. (2004) 'Just leave the dead to rot', *The Guardian*, London, 25 March

Snedaker, S. C. and Snedaker, J. G. (eds) (1984) *The Mangrove Ecosystem: Research Methods*, UNESCO, Paris

Sochaczewski, P. (1999) 'Life reserves: Opportunities to use cultural/spiritual/religious values and partnerships in forest conservation', in Stolton, S., Dudley, N., Gujja, B., Jackson, W. J., Jeanrenaud, J.-P., Oviedo, G. O., Rosabal, P., Phillips, A. and Wells, S. (eds) *Partnerships for Protection*, Earthscan, London

Söderström, L. (1988) 'The occurrence of epixylic bryophyte and lichen species in a natural and a managed forest stand in northwest Sweden', *Biological Conservation*, vol 45, pp169–178

Solberg, B., Brooks, D., Pajuoja, H., Peck, T. J. and Wardle, P. A. (1996) *Long-Term Trends and Prospects in World Supply and Demand for Wood and Implications for Sustainable Forest Management*, European Forest Institute and Norwegian Forest Research Institute, Joensuu, Finland and Høgskoleveinen, Norway

Sollander, E. (2000) *European Forest Scorecards*, WWF International, Gland

Soule, M. and Noss, R. (1998) 'Rewilding and biodiversity: Complementary goals for continental conservation', *Wild Earth*, Spring 1999

Spies, T. A., Franklin, J. F. and Thomas, T. B. (1988) 'Coarse woody debris in Douglas-fir forests of western Oregon and Washington', *Ecology*, vol 69, pp1689–1702

Spurr, S. H. (1952) *Forest Inventory*, The Ronald Press Company, New York

Stattersfield, A. J., Crosby, M. J., Long, A. J. and Wege, D. C. (1998) *Endemic Bird Areas of the World: Priorities for Biodiversity Conservation*, Birdlife International, Cambridge

Steinbeck, J. (1939) *The Grapes of Wrath*, Viking Penguin, New York

Stolton, S. and Dudley, N. (1994) *Consumption and Biodiversity*, a report to WWF International, Equilibrium, Bristol

Stolton, S., Hockings, M., Dudley, N., MacKinnon, K. and Whitten, T. (2003) *Reporting Progress in Protected Areas: A Site-level Management Effectiveness Tracking Tool*, World Bank/WWF Alliance for Forest Conservation and Sustainable Use, Washington DC and Gland, Switzerland

Swanson, F. J. and Dryness, C. T. (1975) 'Impact of clear-cutting and road construction on soil erosion by landslides in the western Cascade Range, Oregon', *Geology*, vol 3, pp393–396

Swanson, F. J., Franklin, J. F. and Sedell, J. (1990) 'Landscape patterns, disturbance and management in the Pacific Northwest, USA', in Zonneveld, I. S. and Forman, R. T. T. (eds) *Changing Landscapes: An Ecological Perspective*, Springer-Verlag, New York

Tarasofsky, R. (1996) *The International Forests Regime*, WWF and IUCN, Gland, Switzerland

Tarasofsky, R. (ed.) (1999) *Assessing the International Forest Regime*, IUCN Environmental Law Centre, Bonn

Taylor, P. (2005) *Beyond Conservation: A Wildland Strategy*, Earthscan, London

Thompson, J. N. (1980) 'Treefalls and colonization patterns of temperate forest herbs', *American Midland Naturalist*, vol 104, pp176–184

Tickle, A., Fergusson, M. and Drucker, G. (1995) *Acid Rain and Nature Conservation in Europe*, WWF, Gland, Switzerland

Tjernberg, M. (1986) *The Golden Eagle and Forestry*, Department of Wildlife Ecology, Report no 12, Swedish University of Agricultural Science, Uppsala

UN (1993) *Agenda 21*, United Nations, New York

UNECE (1988) *Conclusions and Draft Recommendations of the Workshops on Critical Levels for Forests, Crops and Materials and on Critical Loads for Sulphur and Nitrogen*, EB.AIR/R.30, UNECE, Geneva

UNECE and FAO (1994) *Les Ressources Forestières des Zones Tempérées: Analyse des Ressources Forestières de la CEE-ONU/FAO de 1990: Volume 2: Rôle et Fonctions de la Fôret*, UNECE and FAO, New York and Geneva

UNEP (1997) 'Recommendations for a core set of indicators of biological diversity', background paper prepared by the liaison group on indicators of biological diversity, Convention on Biological Diversity, Subsidiary Body of Scientific, Technical and Technological Advice, Third Meeting, Montreal, Canada, 1–5 September, UNEP/CBD/SBSTTA/3/Inf1.3, 22 July, UNEP

UNEP (2004) *Expanded Programme of Work on Forest Biological Diversity*, Convention on Biological Diversity, UNEP, Montreal

UNESC (1994) *Les Ressources Foresteières de la CEE-ONU/FAO de 1990: Volume 2: Rôle et Fonctions de la Fôret*, UNECE and FAO, New York and Geneva

Utting, P. (1991) *The Social Origins and Impact of Deforestation in Central America*, UNRISD Discussion Paper 24, United Nations Research Institute for Social Development, Geneva

Vallauri, D., André, J. and Blondel, J. (2003) *Le Bois Mort: Un Attribute Vital de la Biodiversité de la Forêt Naturelle, une Lacune des Forêts Gérées*, (in French with English translation), WWF France, Paris

Vallauri, D., André, J., Dodelin, B., Eynard-Machet R. and Rambaud, D. (eds) (2005) *Bois Mort et à Cavités*, Editions TEC et Doc, Paris

van Buren, E., Lammerts, M. and Blom, E. M. (1997) *Hierarchical Framework for the Formulation of Sustainable Forest Management Standards: Principles and Criteria*, Tropenbos Foundation, Leiden, The Netherlands

Viana, V. M., Ervin, J., Donovan, R. Z., Elliott, C. and Gholz, H. (1996) *Certification of Forest Products: Issues and Perspectives*, Island Press, Washington DC

Victor, D. G. and Ausubel, J. H. (2000) 'Restoring the forests', *Foreign Affairs*, vol 79, no 6, pp127–144

Virkkala, R. (1987) 'Effects of forest management on birds breeding in northern Finland', *Annales Zoologici Fennica*, vol 24, pp281–294

Ward, J. S. and Parker, G. R. (1989) 'Spatial distribution of woody regeneration in an old-growth forest', *Ecology*, vol 70, pp1279–1285

Warner, C. (1995) *Selecting Tree Species on the Basis of Community Needs*, Community Forestry Field Manual, FAO, Rome

Watt, A. S. (1925) 'On the ecology of British beechwoods with specific reference to their regeneration', *Journal of Ecology*, vol 13, pp27–73

WDPA) (2005) *2005 World Database on Protected Areas*, (compact disc), World Database on Protected Areas, Washington DC

Welten, M. and Sutter, H. C. R. (1982) *Verbreitungsatlas der Farn – und Blütenpflanzen der Schweiz*, www.waldwissen.net/themen/waldoekologie/biodiversitaet/wsl_swiss_webflora_DE

Whitmore, T. C. (1990) *An Introduction to Tropical Rain Forests*, Clarendon Press, Oxford

Whitmore, T. C. and Sayer, J. A. (eds) (1992) *Tropical Deforestation and Species Extinction*, Chapman and Hall and IUCN, London

Wollenberg, E. with Edmunds, D. and Buck, L. (2000) *Anticipating Change: Scenarios as a Tool for Adaptive Forest Management*, Center for International Forestry Research, Bogor, Indonesia, www.cifor.cgiar.org/scripts/newscripts/publications/default.asp

Wong, J., Thornber, K. and Baker, N. (2001) *Resource Assessment of Non-Wood Forest Products: Experience and Biometric Principles*, FAO, Rome

Wood, A., Stedman-Edwards, P. and Mang, J. (eds) (2000) *The Root Causes of Biodiversity Loss*, Earthscan, London

Worrell, R. and Hampson, A. (1997) 'The influence of some forest operations on the sustainable management of forest soils: A review', *Forestry*, vol 70, no 1, pp61–86

WSL (2003) *Schweizerisches Landesforstinventar 1993-95*, Wissenswertes zum Schweizer Wald, Birmensdorf, www.lfi.ch

WWF (1996a) *Measuring Forest Quality: Criteria and Indicators for Forests: WWF Position Statement*, paper for the Second Meeting of the Subsidiary Body on Scientific, Technical and Technological Advice (SBSTTA) of the Convention on Biological Diversity, 2–6 September, WWF, Montreal

WWF (1996b) *Measuring Forest Quality: A paper for the IPF*, Position statement for the Second Meeting of the Intergovernmental Panel on Forests of the Commission on Sustainable Development, Geneva, 11–22 March, WWF, Gland, Switzerland

WWF and IUCN (1998) *Central Africa Workshop: To Draw Up Proposals for Field-Testing Forest Quality Guidelines and for Regional Forest Development*, Yaoundé, 2–3 March, WWF and IUCN, Gland, Switzerland

WWF and IUCN (2000) *Forests Reborn: A Workshop on Forest Restoration*, Gland, Switzerland, WWF and IUCN

WWF, IUCN and EPFL (1998) *Measuring Forest Quality: A Workshop to Examine Options for Measuring forest Quality at a Landscape Scale in Europe*, Gland, Switzerland, April 1–2, Gland and Lausanne, Switzerland, WWF, IUCN and EPFL

WWF, IUCN, GTZ and EPFL (1999) *Forest Quality/Calidad Forestal/La Qualite des Forêts*, Gland, Switzerland, WWF, IUCN, GTZ and EPFL

Index

Page numbers in *italics* refer to figures, tables and boxes

Printed and bound by CPI Group (UK) Ltd, Croydon, CR0 4YY

22/10/2024

01777611-0001